C0-AOO-009

# ELEMENTARY
# ALGEBRA

# ELEMENTARY ALGEBRA

**Robert D. Hackworth**
**Robert H. Alwin**

*St. Petersburg Junior College*
*Clearwater, Florida*

Holt, Rinehart and Winston, Inc.
*New York   Chicago   San Francisco   Atlanta*
*Dallas   Montreal   Toronto   London   Sydney*

Copyright © 1972 by Holt, Rinehart and Winston, Inc.
All Rights Reserved
Library of Congress Catalog Card Number: 73-145746
**ISBN: 0-03-077820-4**
Printed in the United States of America
2 3 4 5   038   1 2 3 4 5 6 7 8 9

# Preface

Prospective readers of this text, whether they be students or teachers, will probably approach its use with a quizzical attitude as to the possibility that this book might be different or better than the hundreds of other Algebra books on the market. Such an attitude is the first that a prospective author must overcome before embarking on the writing of a text that covers essentially the same material as a great number of already-published books. We, the authors, had to answer the question, "What can we do in authoring a text that is an improvement over all the efforts of our colleagues?" In this preface we hope to explain part of our answer, but the rest must remain with the students and teachers who work diligently through its pages to determine whether we have, in fact, offered an improvement over other Algebra texts. We think we have.

Perhaps our overriding reason for having the temerity to compose such a work is our past teaching experience. Authors of beginning Algebra texts are often acquainted only with mathematics rather than having, in addition, a number of years' experience teaching a beginning Algebra course. Between us we count over thirty-five years experience in teaching algebra to students in junior high schools, high schools, and junior colleges.

That experience is evident in the explanations given throughout the text, our selection of exercise problems, and the arrangement of topics.

Another reason for undertaking the writing of this text was our conclusion that too many students are forced to learn all of their algebra from teacher presentations in class simply because the texts that are used serve as mere outlines for the material along with thousands of problems that would be more appropriate in a workbook. It is our firm conviction that a text is responsible for presenting and explaining its material. In short, we believe that a text must teach its contents, thereby freeing the teacher from the time-consuming task of spoonfeeding information and making him available to provide further explanation of subtle or difficult concepts. Although the text was written for traditional classroom use it is certainly not limited to group instruction. Any student or instructor seeking individualized study materials should find the content and presentation here to be well-suited for that purpose.

One difficulty with many texts on the market is that they were written by mathematics teachers to please their colleagues with the beauty of their mathematical exposition. We feel the best way to please our colleagues is to supply them with a text that best suits the needs of their students for home study and learning. For this reason the explanations and illustrations are written in the hope that the student will find the text to be his teacher at home. The writing is straightforward, presenting good mathematics in the most direct and simple language at our command.

The first seven chapters of the text are a development of the concepts and skills of the algebra of the rational numbers. As intermediate steps of this development, the language of sets is developed in Chapter 1, the algebra of the counting numbers in Chapter 2, the algebra of the integers in Chapter 3, and the structure of the rational number system in Chapter 4. The development of algebras for the counting numbers, integers, and rational numbers has proved to be a very effective way to promote and reinforce the learning process. That is, the basic properties of the real number system can be studied, re-studied, and then studied again, each time in a new context that broadens understanding of the properties of the different number systems.

Chapter 5 deals with linear equations and inequalities with two variables, still using the field of rational numbers as the basis for the algebraic manipulations. Similarly, polynomials, their multiplication, their factoring, and problems involving the simplification, multiplication, and addition of polynomial fractions are explained in Chapter 6. Chapter 7 consists of a review of the equation and inequality-solving skills of earlier chapters while introducing the solution of quadratic equations by factoring. The fact that some quadratic polynomials are prime leads to the conclusion that some polynomial equations have no solution in the set of rational numbers, and this conclusion is used to motivate the development of irrational numbers in Chapter 8.

Besides the development of arithmetic skills for irrational numbers involving radicals, Chapter 8 presents the structure of the real number system and extends the solution of quadratic equations through the use of the quadratic formula.

Chapter 9 deals with the study of absolute value equations and inequalities and Chapter 10 is a presentation of relations and functions.

Each unit of the text is designated by a numeral representing the chapter and the unit's sequencing in that chapter. For example, the unit designated 4.15 is the fifteenth unit of Chapter 4. Whenever a unit seemed desirably split into more than one part extra exercises have been included and the exercises keyed to their particular section. For example, the exercise designated 3.5.2 is in the third chapter, fifth unit, and is the second exercise of that unit.

The authors want to take this opportunity to thank the many people who have aided us in the writing and production of this text. Most appreciated are the hundreds of students we have had the pleasure of working with in the last few years, who have served as our best critics in this endeavor. Also, a word of thanks goes to the National Science Foundation and its programs in teacher education. Not only have we received a portion of our education from the NSF programs, but we owe our acquaintance to each other to a program offered at the University of Arizona in the summers of 1962, 1963, and 1964.

Robert D. Hackworth
Robert H. Alwin

Clearwater, Florida
November, 1971

# Contents

## 3 / The Integers    55

# 4 / The Rational Numbers    97

# 7 / Solving Equations and Inequalities    241

# ELEMENTARY
# ALGEBRA

# 1 SETS, THEIR USE, AND THEIR OPERATIONS

## 1.1 Introduction

Answers for many questions or problems in mathematics depend upon the particular context in which the question or problem is stated. As an example of this statement consider the following story:

Mrs. Wilson is a second grade teacher. She gave her class the following three-problem quiz:

1. Find a pair of numbers that can be multiplied to give a product of 15.
2. What is the greatest number less than 20?
3. Find a number between 6 and 9.

When the class finished the quiz, Mrs. Wilson put the answers on the board.

1. 1 and 15 or 3 and 5.
2. 19
3. 7 or 8

John Jones, one of the pupils, complained bitterly because he felt his answers were correct even though they were different from Mrs. Wilson's. John's answers were as follows:

1. $7\frac{1}{2}$ and 2
2. There is no greatest number less than 20.
3. $8\frac{1}{4}$

Do you think that John should receive credit for his answers?

The problem that developed in Mrs. Wilson's class came about because Mrs. Wilson asked questions within the context of second grade arithmetic while John, being a very unusual second grader, viewed the problems within a much broader number concept. The story may seem to exaggerate the situation, but in mathematics it is extremely important to state the kind of numbers that may be used in working a given problem.

EXERCISE 1.1.1

1. How could Mrs. Wilson have stated the problems so that only her answers would be correct?
2. John often fooled his friends with the following problem: "I am thinking of a number between 1 and 10. I will give you seven guesses to select the number. If you guess it you win; otherwise I win." Why did John always win?

## 1.2   Sets

Stating clearly the numbers or objects to be considered in a problem is accomplished in mathematics by the concept of sets. The use of the word "set" in mathematics is similar to its use in phrases such as "a set of dishes" and "a set of golf clubs." Any collection or group of numbers or objects or names can be called a set. Any collection of symbols, $=$, $>$, $+$, and $\sqrt{}$ could be called a set.

Braces, { }, are used to designate sets. The set of numbers 4, 17, 34, and 63 can be shown as {4, 17, 34, 63}. The set of whole numbers between 6 and 11 can be shown {7, 8, 9, 10}. The set of even numbers greater than 19 and less than 30 can be shown {20, 22, 24, 26, 28}.

EXERCISE 1.2.1

Use braces to show each of the following sets:

1. The set of whole numbers 6, 19, 104, and 817.
2. The set of symbols @, %, and #.
3. The set of whole numbers greater than 16 and less than 23.
4. The set of even numbers between 73 and 81.

5. The set of numbers between 7 and 29 that are divisible by 3.
6. The set of odd numbers greater than 19 and less than 28.
7. The set of whole numbers greater than 71 and less than 80.
8. The set of numbers between 3 and 27 that are divisible by 5.
9. The set of whole numbers greater than 19 and less than 24.
10. The set of odd numbers between 8 and 12.
11. The set of whole numbers 17, 23, 96 and 43.
12. The set of symbols $>$, $\leq$, $=$ and $\sqrt{\phantom{x}}$.
13. The set of whole numbers greater than 12 and less than 19.
14. The set of even numbers between 125 and 135.
15. The set of numbers between 8 and 28 that are divisible by 9.
16. The set of odd numbers greater than 31 and less than 42.
17. The set of whole numbers greater than 57 and less than 68.
18. The set of numbers between 43 and 65 that are divisible by 8.
19. The set of whole numbers greater than 25 and less than 30.
20. The set of odd numbers between 33 and 45.

The set of all whole numbers greater than 11 and less than 93 can be shown as {12, 13, 14, ..., 92}. The three dots used in {12, 13, 14, ..., 92} are to be read "and so forth" and indicate that the reader should keep counting as indicated by the first three numbers in the set until he reaches 92. In the same way, the set of all even numbers greater than 15 and less than 117 can be shown as {16, 18, 20, ..., 116}. In this set the three dots indicate that the pattern of even numbers is to continue on to 116.

The set {5, 6, 7, ...} does not have a number following the three dots. This symbolization means that the counting is to go on forever and there is no highest number in the set. {5, 6, 7, ...} is the set of all whole numbers greater than 4. {13, 15, 17, ...} is the set of all odd numbers greater than 12.

EXERCISE 1.2.2

Use braces and three dots to properly show each of the following sets:

1. The set of whole numbers greater than 9 and less than 47.
2. The set of whole numbers greater than 47.
3. The set of odd numbers greater than 48 and less than 417.
4. The set of whole numbers greater than 5 and less than 99 that can be evenly divided by 4.
5. The set of all odd numbers greater than 812.
6. The set of all whole numbers greater than 2 that can be evenly divided by 5.
7. The set of all whole numbers greater than 90.
8. The set of all whole numbers less than 57.

9. The set of all odd numbers between 40 and 81.
10. The set of all even numbers between 103 and 201.
11. The set of whole numbers greater than 14 and less than 61.
12. The set of whole numbers greater than 901.
13. The set of odd numbers greater than 93 and less than 401.
14. The set of whole numbers greater than 8 and less than 123 that can be evenly divided by 10.
15. The set of all odd numbers greater than 504.
16. The set of all whole numbers greater than 6 that can be evenly divided by 7.
17. The set of all whole numbers greater than 66.
18. The set of all whole numbers less than 88.
19. The set of all odd numbers between 35 and 136.
20. The set of all even numbers between 240 and 375.

A set is a group or collection of numbers or objects. {9, 13, 18} is a set of numbers. {+, =, √, <} is a set of mathematical symbols. The set itself is a group or collection consisting of certain members, called "elements" of the set. 9 is in the set {1, 8, 9, 12} so we state that, "9 is an element of the set." 23 is *not* in the set {9, 13, 18} so we state that "23 is *not* an element of the set." "∈" is a symbol for "is an element of" and "∉" is a symbol for "*not* an element of." 9 ∈ {1, 8, 9, 12} is a symbolization of the statement, "9 is an element of the set {1, 8, 9, 12}. 23 ∉ {9, 13, 18} is a symbolization of the statement, "23 is *not* an element of the set {9, 13, 18}."

43 ∈ {16, 27, 43, 96, 107} is a true statement because 43 is an element of the set {16, 27, 43, 96, 107}.

16 ∉ {2, 3, 9, 17, 43, 903} is true because 16 is not an element of the set {2, 3, 9, 17, 43, 903}.

19 ∈ {4, 5, 6, . . ., 47} is a true statement because the three dots indicate that all whole numbers between 3 and 48 are elements of the set.

14 ∈ {9, 11, 13, . . ., 93} is a false statement because the three dots indicate that all odd numbers between 8 and 94 are elements of the set. Since 14 is not an odd number, 14 ∉ {9, 11, 13, . . ., 93} is a true statement.

EXERCISE 1.2.3

Determine whether each of the following statements is true or false:

1. 4 is an element of the set {3, 4, 7, 10}.
2. 8 is not an element of the set {4, 5, 7, 8, 10}.
3. 3 ∈ {1, 4, 11, 12, 15}.
4. 9 ∉ {2, 5, 8, 12}.
5. 47 ∈ {1, 2, 3, . . .}.

6.  26 ∈ {8, 9, 10, ..., 24}.
7.  192 ∈ {2, 4, 6, ...}.
8.  23 ∉ {5, 10, 15, ..., 45}.
9.  263 ∈ {150, 151, 152, ...}.
10. 63 ∉ {1, 3, 5, ...}.
11. 17 is an element of {2, 10, 15, 16, 18}
12. 9 is not an element of {5, 7, 9, 11, 13}
13. 8 ∈ {2, 5, 7, 8, 15}
14. 16 ∉ {5, 10, 16, 20}
15. 96 ∈ {1, 2, 3, ...}
16. 79 ∈ {43, 45, 47, ..., 69}
17. 822 ∈ {2, 4, 6, ...}
18. 26 ∉ {4, 8, 12, ..., 44}
19. 915 ∈ {71, 72, 73, ...}
20. 576 ∉ {2, 4, 6, ...}

## 1.3  Special Sets

A few special sets deserve mention here. {1, 2, 3, ...} is called the *set of counting numbers* because it includes all numbers that could be used to count objects. Zero is not an element of {1, 2, 3, ...}, because zero is not a counting number. There must be at least one object to start the counting process. The importance of the set of counting numbers is best emphasized by the reminder that the set {1, 2, 3, ...} contains the first numbers studied in elementary school and these numbers form the basis for all arithmetic generally presented through the elementary grades.

A second important set is called the *empty set* and is designated by { }. The empty set has no elements. The blank space between the braces in { } aptly describes the empty set. The student may wonder why the empty set is considered important when in fact it has no elements. It is important because there are many situations in mathematics which have no number answers. There is *no counting number* that can be added to 7 to give 3. (7 + __ = 3). There is *no counting number* that can be multiplied by 5 to give a product of 13. (5·__ = 13). Hence the set of counting numbers that can be added to 7 to give 3 is { }. Also, the set of counting numbers that can be multiplied by 5 to give 13 is { }.

A third set which deserves mention is called a *singleton (or unit) set*. {7}, {16}, {93}, and {403} are examples of singleton sets because each contains exactly one element. The importance of singleton sets lies in the fact that many mathematical problems present situations in which there is only one true response. The set of all counting numbers which can be added to 9 to give 16 is {7}, because 7 is the only counting number with the desired property.

EXERCISE 1.3.1

Determine whether each of the following is the set of counting numbers, the empty set, or a singleton set:

1.  The set of all numbers greater than 15 and also less than 4.
2.  The set of all numbers which can be multiplied by 3 to give a product of 21.
3.  The set of all whole numbers greater than zero.
4.  The set of all whole numbers which can be multiplied by 6 to give a product of 10.
5.  The set of all whole numbers that can be added to 14 to give a sum of 19.
6.  The set of all whole numbers that can be evenly divided by 1.
7.  The set of all counting numbers greater than 16 or less than 20.
8.  The set of all numbers that can be added to 15 to give 30.
9.  The set of all numbers that can be multiplied by 7 to give 42.
10. The set of all odd counting numbers between 8 and 11.

## 1.4   Union of Sets

$\{6, 10, 11\}$ and $\{4, 10, 15, 47\}$ are two sets which have counting numbers as elements. A third set can be obtained from $\{6, 10, 11\}$ and $\{4, 10, 15, 47\}$ by taking *all* the different elements and placing them in a common set. This third set would be $\{4, 6, 10, 11, 15, 47\}$, and is the "union" of the sets $\{6, 10, 11\}$ and $\{4, 10, 15, 47\}$. "∪" is the symbol to designate union. Using the union symbol, "∪," the above example may be shown as follows:

$$\{6, 10, 11\} \cup \{4, 10, 15, 47\} = \{4, 6, 10, 11, 15, 47\}$$

Study very carefully the following examples which illustrate the union of two sets. The union of two sets is the set which consists of *all* those elements found in either or both of the two original sets. An element in both sets being unioned appears only once in the resulting set;

$$\{1, 4, 5\} \cup \{4, 5, 7, 9\} = \{1, 4, 5, 7, 9\}$$

$$\{8, 14, 19, 20\} \cup \{3, 8, 9, 37\} = \{3, 8, 9, 14, 19, 20, 37\}$$

$$\{1, 2, 3, \ldots, 51\} \cup \{5, 21, 52\} = \{1, 2, 3, \ldots, 52\}$$

$$\{4, 13, 22\} \cup \{1, 4, 7, 13, 17, 19, 22\} = \{1, 4, 7, 13, 17, 19, 22\}$$

$$\{2, 4, 7\} \cup \{\quad\} = \{2, 4, 7\}$$

$$\{1, 2, 3, \ldots, 97\} \cup \{51, 52, 53, \ldots, 131\} = \{1, 2, 3, \ldots, 131\}$$
$$\{2, 4, 6, \ldots\} \cup \{4, 8, 12, \ldots\} = \{2, 4, 6, \ldots\}$$
$$\{\ \ \} \cup \{5, 9\} = \{5, 9\}$$
$$\{4, 7, 39\} \cup \{1, 2, 3, \ldots\} = \{1, 2, 3, \ldots\}$$

EXERCISE 1.4.1

1.  $\{6, 15, 21\} \cup \{5, 6, 7, 19\} =$
2.  $\{2, 3, 4\} \cup \{4, 5, 6\} =$
3.  $\{4, 7, 9, 13, 15, 18\} \cup \{2, 6, 10\} =$
4.  $\{3, 4, 5, 73\} \cup \{4\} =$
5.  $\{2\} \cup \{97\} =$
6.  $\{4, 7, 13, 16\} \cup \{\ \ \} =$
7.  $\{\ \ \} \cup \{6, 93, 107\} =$
8.  $\{1, 2, 3, \ldots, 91\} \cup \{6, 9, 92, 93\} =$
9.  $\{71, 72, 73, \ldots, 104\} \cup \{1, 2, 3, \ldots\} =$
10.  $\{5, 10, 15, \ldots\} \cup \{10, 20, 30, \ldots\} =$
11.  $\{3, 5, 7\} \cup \{2, 6, 7, 9, 11\} =$
12.  $\{3, 8\} \cup \{5, 10, 15, 16\} =$
13.  $\{2, 9, 10, 11, 12\} \cup \{3, 4, 9, 10\} =$
14.  $\{15, 16, 17\} \cup \{17\} =$
15.  $\{1, 2, 3, \ldots, 15\} \cup \{10, 11, 12, \ldots, 20\} =$
16.  $\{5, 10, 15, \ldots\} \cup \{\ \ \} =$
17.  $\{\ \ \} \cup \{1, 2, 3, \ldots, 10\} =$
18.  $\{47, 49, 51, \ldots, 95\} \cup \{1, 3, 5, \ldots\} =$
19.  $\{2, 57, 75, 84\} \cup \{1, 2, 3, \ldots, 83\} =$
20.  $\{6, 12, 18, \ldots\} \cup \{3, 6, 9, \ldots\} =$

$\{6, 15, 27, 30\}$ and $\{5, 10, 15, 20\}$ are two sets of counting numbers. When the two sets are unioned by taking $\{6, 15, 27, 30\}$ as the first set and $\{5, 10, 15, 20\}$ as the second set the result is $\{5, 6, 10, 15, 20, 27, 30\}$. Exactly the same result is obtained if $\{5, 10, 15, 20\}$ is used as the first set and $\{6, 15, 27, 30\}$ is the second set. The fact that the union of two sets is the same regardless of the order in which the operation is performed may not be surprising, but it is of sufficient importance in mathematics to state it as a property as follows:

*The Commutative Law for the Union of Two Sets*

The result of unioning two sets is the same regardless of the order in which they are taken.

$$\{1, 2, 3, 4\} \cup \{2, 5\} = \{2, 5\} \cup \{1, 2, 3, 4\}$$

When the empty set, { }, is unioned with a second set the result is relatively easy to obtain because it will always be the second set. For example, each of the following union problems has been completed correctly:

$$\{ \ \} \cup \{5, 6, 7\} = \{5, 6, 7\}$$

$$\{ \ \} \cup \{1, 2, 3, \ldots, 813\} = \{1, 2, 3, \ldots, 813\}$$

$$\{ \ \} \cup \{7, 14, 21, \ldots\} = \{7, 14, 21, \ldots\}$$

The empty set is unique in its behavior within a union problem because the result of unioning any set with the empty set is always identical with the other set. For this reason the empty set is called the Identity set for the operation union.

*The Empty Set is the Identity Set for Union of Sets*

> Whenever the empty set is unioned with a second set the result is the second set.
>
> $\{ \ \} \cup \{2, 3\} = \{2, 3\}$

Union is an operation performed on two sets. It is possible to union more than two sets by performing the operation union with two sets at a time. For example, $\{6, 9\} \cup \{7, 10, 11\} \cup \{6, 7, 8\}$ is a union problem in which the operation union must be performed on more than two sets. Since union can only be performed on two sets at a time, parentheses are used to show which two sets are to be unioned first, as follows:

$$\{6, 9\} \cup \{7, 10, 11\} \cup \{6, 7, 8\}$$

$$(\{6, 9\} \cup \{7, 10, 11\}) \cup \{6, 7, 8\}$$

$$\{6, 7, 9, 10, 11\} \cup \{6, 7, 8\}$$

$$\{6, 7, 8, 9, 10, 11\}$$

The parentheses in $\{4, 9, 13\} \cup (\{6\} \cup \{8, 9, 13\})$ are placed around the last two sets and indicate that they are to be unioned first. The problem is done as follows:

$$\{4, 9, 13\} \cup (\{6\} \cup \{8, 9, 13\})$$

$$\{4, 9, 13\} \cup \{6, 8, 9, 13\}$$

$$\{4, 6, 8, 9, 13\}$$

The union of sets is accomplished on only two sets at a time. Whenever three sets are to be unioned it is necessary to show which two sets are to be unioned first and this is accomplished by the use of parentheses. Whenever

no parentheses are in a statement of union, the operation union should be
performed on the first two sets, as shown in the following example:

$$\{1, 2, 3\} \cup \{2, 5, 7\} \cup \{3, 5, 8\}$$

$$(\{1, 2, 3\} \cup \{2, 5, 7\}) \cup \{3, 5, 8\}$$

$$\{1, 2, 3, 5, 7\} \cup \{3, 5, 8\}$$

$$\{1, 2, 3, 5, 7, 8\}$$

EXERCISE 1.4.2

1.   $(\{4, 21, 37\} \cup \{5, 21, 24\}) \cup \{4, 19\} = $
2.   $\{2, 7, 13, 19\} \cup (\{3, 5\} \cup \{2, 5\}) = $
3.   $(\{6, 12, 18, 20\} \cup \{96, 104, 168\}) \cup \{2, 4, 6, \ldots\} = $
4.   $\{4, 21, 37\} \cup (\{5, 21, 24\} \cup \{4, 19\}) = $
5.   $(\{2, 7, 13, 17\} \cup \{3, 5\}) \cup \{2, 6\} = $
6.   $\{6, 12, 18, 20\} \cup (\{96, 104, 168\} \cup \{2, 4, 6, \ldots\}) = $
7.   $\{ \ \ \} \cup (\{3, 4, 10\} \cup \{4, 6, 10\}) = $
8.   $(\{1, 2, 3, \ldots\} \cup \{ \ \ \}) \cup \{5, 10, 68\} = $
9.   $\{2, 4, 5\} \cup \{4, 5, 10\} \cup \{ \ \ \} = $
10.  $\{7, 12, 19, 21\} \cup \{ \ \ \} \cup \{19, 22\} = $

From the results of the preceding exercise it may be surmised that the
union of three sets gives the same result regardless of the placement of the
parentheses. This is a true observation and its statement as a property of
union follows:

*The Associative Law for the Union of Sets*

In unioning three sets the result is the same whether parentheses
are used to group the first two sets or the last two sets.

EXERCISE 1.4.3

For each of the following equalities state the property of union expressed by the statement:

1.   $\{ \ \ \} \cup \{6, 9, 14, 107, 206\} = \{6, 9, 14, 107, 206\}$
2.   $\{84, 93, 168, 512\} \cup \{19, 37, 406\}$
         $= \{19, 37, 406\} \cup \{84, 93, 168, 512\}$
3.   $\{16, 19, 20\} \cup (\{3, 4\} \cup \{4, 19, 25\})$
         $= (\{16, 19, 20\} \cup \{3, 4\}) \cup \{4, 19, 25\}$
4.   $\{ \ \ \} \cup \{3, 6, 9, \ldots\} = \{3, 6, 9, \ldots\} \cup \{ \ \ \}$

5.   $\{5, 17, 33\} \cup \{1, 2, 3, \ldots, 32\} = \{1, 2, 3, \ldots, 32\} \cup \{5, 17, 33\}$
6.   $(\{6, 41\} \cup \{1, 2, 3, \ldots, 10\}) \cup \{19, 67\}$
$$= \{6, 41\} \cup (\{1, 2, 3, \ldots, 10\} \cup \{19, 67\})$$
7.   $\{5, 10, 15, \ldots\} = \{\ \ \} \cup \{5, 10, 15, \ldots\}$
8.   $\{2, 8, 15\} \cup \{3, 8, 10\} = \{3, 8, 10\} \cup \{2, 8, 15\}$
9.   $\{4, 8, 9\} \cup (\{2, 6\} \cup \{\ \ \}) = (\{4, 8, 9\} \cup \{2, 6\}) \cup \{\ \ \}$
10.  $\{1, 2, 3, \ldots, 10\} \cup \{3, 6, 9\} = \{3, 6, 9\} \cup \{1, 2, 3, \ldots, 10\}$

### 1.5   Intersection of Sets

A third set can be formed from $\{5, 13, 19, 21\}$ and $\{6, 13, 19, 47\}$ by taking all those elements that appear in *both* sets. Since the sets $\{5, 13, 19, 21\}$ and $\{6, 13, 19, 47\}$ both have 13 and 19 as elements, the third set would be $\{13, 19\}$. Whenever two sets are used to find a third set containing all the elements that are common to both sets the operation is called intersection. "$\cap$" is the symbol for intersection. The example above may be shown by using the symbol "$\cap$" as follows:

$$\{5, 13, 19, 21\} \cap \{6, 13, 19, 47\} = \{13, 19\}$$

Below are a number of examples illustrating the intersection of two sets. Study these examples carefully.

$$\{2, 9, 33, 49\} \cap \{4, 7, 9, 36, 49\} = \{9, 49\}$$

$$\{3, 6, 9, 12, 15\} \cap \{1, 2, 3, \ldots, 12\} = \{3, 6, 9, 12\}$$

$$\{1, 2, 3, \ldots, 47\} \cap \{30, 31, 32, \ldots, 101\} = \{30, 31, 32, \ldots, 47\}$$

$$\{\ \ \} \cap \{7, 9, 11, 15, 21, 22\} = \{\ \ \}$$

$$\{14, 21, 36, 50\} \cap \{8, 16, 24, \ldots\} = \{\ \ \}$$

$$\{6, 9, 13\} \cap \{3, 5, 6, 8, 9, 10, 13, 15\} = \{6, 9, 13\}$$

$$\{2, 4, 6, \ldots\} \cap \{3, 6, 9, \ldots\} = \{6, 12, 18, \ldots\}$$

Note from the above illustrations that the intersection of two sets consists of only those elements which are common to both sets.

EXERCISE 1.5.1

1.   $\{4, 8, 9, 17\} \cap \{2, 6, 8, 9, 10\} =$
2.   $\{1, 2, 3, \ldots, 10\} \cap \{3, 5, 6, 11, 71, 83\} =$
3.   $\{2, 4, 6, \ldots, 46\} \cap \{9, 12, 26, 33, 40\} =$
4.   $\{5, 10, 13, 89\} \cap \{11, 74, 98\} =$
5.   $\{4, 17, 53, 55, 56\} \cap \{53\} =$

6.  $\{ \quad \} \cap \{41, 47, 103\} =$
7.  $\{1, 2, 3, \ldots, 25\} \cap \{13, 14, 15, \ldots, 802\} =$
8.  $\{1, 2, 3, \ldots, 39\} \cap \{3, 6, 9, \ldots\} =$
9.  $\{1, 2, 3, \ldots\} \cap \{5, 12, 31, 46\} =$
10. $\{4, 8, 12, \ldots\} \cap \{5, 10, 15, \ldots\} =$
11. $\{3, 10, 11\} \cap \{1, 2, 3, \ldots, 10\} =$
12. $\{2, 4, 6, \ldots, 20\} \cap \{7, 10, 12, 15, 16, 22\} =$
13. $\{2, 9, 15\} \cap \{3, 6\} =$
14. $\{3, 6, 9, 12\} \cap \{1, 9, 15\} =$
15. $\{3, 10, 15, 19\} \cap \{1, 2, 3, \ldots, 20\} =$
16. $\{6, 12, 18, \ldots\} \cap \{2, 4, 6, \ldots\} =$
17. $\{1, 2, 3, \ldots, 512\} \cap \{513, 514, 515, \ldots\} =$
18. $\{2, 4, 6, \ldots, 80\} \cap \{70, 72, 74, \ldots\} =$
19. $\{1, 2, 3, \ldots\} \cap \{5, 6, 11, 17\} =$
20. $\{3, 6, 9, \ldots\} \cap \{5, 10, 15, \ldots\} =$

As was the case for union, the result of intersecting two sets is the same regardless of which set is taken first. For example:

$$\{5, 73, 96, 107\} \cap \{2, 4, 5, 19, 96, 100\} = \{5, 96\}$$

and

$$\{2, 4, 5, 19, 96, 100\} \cap \{5, 73, 96, 107\} = \{5, 96\}$$

*The Commutative Law for the Intersection of Two Sets*

The result of intersecting two sets is the same regardless of the order in which they are taken.

Intersection is an operation to be performed on only two sets at one time. Whenever three sets are to be intersected parentheses are used to indicate the two sets that are to be intersected first, as indicated in the following example:

$$(\{5, 11, 43, 96, 103\} \cap \{1, 2, 3, \ldots, 80\}) \cap \{2, 5, 29, 43\}$$

$$\{5, 11, 43\} \cap \{2, 5, 29, 43\}$$

$$\{5, 43\}$$

EXERCISE 1.5.2

1.  $(\{2, 5, 6, 7, 9\} \cap \{1, 2, 3, \ldots, 8\}) \cap \{4, 5, 6, 7\} =$
2.  $\{1, 2, 3, \ldots, 27\} \cap (\{5, 10, 15, \ldots, 50\} \cap \{18, 19, 20, \ldots, 84\}) =$
3.  $(\{2, 7, 12, 17, 22, 27\} \cap \{1, 2, 3, \ldots\}) \cap \{2, 4, 6, \ldots\} =$
4.  $\{2, 5, 6, 7, 9\} \cap (\{1, 2, 3, \ldots, 8\} \cap \{4, 5, 6, 7\}) =$

5.   $(\{1, 2, 3, \ldots, 27\} \cap \{5, 10, 15, \ldots, 50\}) \cap \{18, 19, 20, \ldots, 84\} =$

6.   $\{2, 7, 12, 17, 22, 27\} \cap (\{1, 2, 3, \ldots\} \cap \{2, 4, 6, \ldots\}) =$

7.   $(\{3, 6, 8\} \cap \{\phantom{xx}\}) \cap \{5, 8, 11, 14\} =$

8.   $\{2, 5, 9, 10\} \cap (\{2, 4, 9, 15, 19\} \cap \{\phantom{xx}\}) =$

9.   $(\{4, 12, 19\} \cap \{3, 6, 9, \ldots\}) \cap \{1, 2, 3, \ldots\} =$

10.  $(\{5, 10, 15, \ldots\} \cap \{2, 4, 6, \ldots\}) \cap \{10, 20, 30, \ldots\} =$

Although intersection can only be performed on two sets at a time the result of intersecting three sets is the same regardless of the placement of the parentheses.

## *The Associative Law for the Intersection of Sets*

In intersecting three sets the result is the same whether parentheses are used to group the first two sets or the last two sets.

### 1.6   Ordered Pairs

One other type of operation on sets is to be discussed in this chapter, but before its introduction the concept of ordered pair must be presented.

Each of the sets $\{3, 5\}$ and $\{5, 3\}$ has two elements, and though the elements are listed in different orders, the sets have exactly the same elements. $\{3, 5\}$ and $\{5, 3\}$ are equal sets because the order of listing makes no difference in describing the group of numbers which are elements of the set. For the same reason, $\{2, 17\}$ is equal to $\{17, 2\}$ and $\{19, 407\}$ is equal to $\{407, 19\}$.

$(4, 3)$ is an ordered pair of real numbers. Unlike a set of two elements the order of the numbers in an ordered pair is important. $(4, 3)$ is not equal to $(3, 4)$ because the concept of ordered pair requires the listing of the numbers in a particular order. $(2, 9)$ is not equal to $(9, 2)$ because the numbers 2 and 9 are in different orders. For the same reason $(6, 13)$ is not equal to $(13, 6)$ and $(41, 97)$ is not equal to $(97, 41)$.

In the ordered pair $(12, 35)$, 12 is called the first component and 35 is called the second component. Similarly, 7 is the first component of $(7, 15)$ and 15 is the second component.

Ordered pairs can be elements of a set. For example,

$$\{(2, 7), (3, 9), (4, 17)\}$$

is a set with three elements and each element is an ordered pair. $\{(9, 16)\}$ is a singleton set; it has only one element which is the ordered pair $(9, 16)$. The individual components of an ordered pair are not elements of the set of ordered pairs.

EXERCISE 1.6.1

True or false:

1.  The first component of (5, 16) is 5.
2.  The second component of (8, 47) is 8.
3.  There are four elements in the set, {(2, 7), (8, 9)}.
4.  There are three elements in the set, {(5, 6), (2, 7), (8, 13)}.
5.  (2, 3) ∈ {(5, 4), (6, 1), (3, 2), (8, 1)}.
6.  (6, 9) ∈ {(4, 7), (3, 8), (9, 1), (6, 9), (5, 4)}.
7.  (2, 6) ∉ {(5, 1), (2, 6), (8, 3)}.
8.  (4, 9) ∉ {(2, 7), (9, 4), (6, 1), (2, 5)}.
9.  6 ∈ {(8, 4), (2, 6), (6, 7)}.
10.  9 ∉ {(9, 9), (9, 7)}.
11.  The first component of (7, 23) is 30.
12.  The second component of (5, 8) is 8.
13.  There are three elements in {(2, 4), (6, 8), (10, 12)}.
14.  There are four elements in {(7, 3), (2, 2)}.
15.  (6, 7) ∈ {(4, 1), (5, 7), (6, 7), (8, 3)}.
16.  (3, 5) ∉ {3, 4, 5, 6}.
17.  (2, 8) ∈ {(1, 5), (2, 4), (3, 3)}.
18.  (5, 11) ∉ {(1, 4), (3, 7), (11, 5), (13, 2)}.
19.  4 ∈ {(3, 1), (4, 4), (5, 4), (1, 6)}.
20.  6 ∉ {(4, 2), (3, 3), (1, 5)}.

## 1.7   The Cartesian Product of Two Sets

The sets {2, 3} and {5, 6, 7} can be used to generate a third set called the Cartesian Product of {2, 3} and {5, 6, 7}. The Cartesian Product of {2, 3} and {5, 6, 7} is {(2, 5), (2, 6), (2, 7), (3, 5), (3, 6), (3, 7)}. The ordered pairs are obtained by using the elements of {2, 3} as first components and the elements of {5, 6, 7} as second components. "✕" is the symbol for Cartesian Product

$$\{2, 3\} \times \{5, 6, 7\} = \{(2, 5), (2, 6), (2, 7), (3, 5), (3, 6), (3, 7)\}$$

As another example, the Cartesian Product of {6, 9, 10, 13} and {4, 7} follows. Notice that the ordered pairs in the set have first components from {6, 9, 10, 13} and the second components from {4, 7}.

{6, 9, 10, 13} ✕ {4, 7}

$$= \{(6, 4), (6, 7), (9, 4), (9, 7), (10, 4), (10, 7), (13, 4), (13, 7)\}$$

EXERCISE 1.7.1

Find the following Cartesian Products:

1.  {6, 9} × {8, 10, 11}
    There will be six elements (ordered pairs) in the Cartesian Product. Why?
2.  {1, 5, 9} × {2, 6, 8}
    How many elements are there in the set?
3.  {3, 6, 9, 10} × {1, 5}
4.  {2, 3, 4, 5} × {6, 7, 8}
5.  {7} × {3, 5, 6, 9, 10}
6.  {2, 7, 9} × {6}
7.  {4, 5} × {4, 6, 8}
8.  {3, 4, 5} × {3, 4, 5}
9.  {5} × {7}
10. {6, 3, 4, 8} × {   }

The Cartesian Product of two sets is not a commutative operation. The order in which two distinct sets are taken usually will give different results, as shown in the following example:

{2, 3} × {1, 6, 9} = {(2, 1), (2, 6), (2, 9), (3, 1), (3, 6), (3, 9)}

{1, 6, 9} × {2, 3} = {(1, 2), (1, 3), (6, 2), (6, 3), (9, 2), (9, 3)}

Notice that the Cartesian Products do not have exactly the same ordered pairs as elements. (2, 1) is an element of {2, 3} × {1, 6, 9}, but (2, 1) is not an element of {1, 6, 9} × {2, 3}.

### 1.8  Graphing

The set of all counting numbers can be represented pictorially by a half-line as in Figure 1.1. The arrow at the right side of the line indicates

FIGURE 1.1

that the half-line continues forever to the right. No arrow is needed at the left side of the drawing because 1 is the least counting number.

The set {3, 4, 7} may be graphed on the half-line by making a visible dot at the position on the half-line to indicate the numbers 3, 4, and 7. This graph would be as follows:

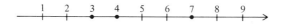

FIGURE 1.2

EXERCISE 1.8.1

Graph each of the following sets on a half-line.

1. {1, 5, 8}
2. {2, 3, 4, 6, 9}
3. {1, 6, 9, 11}
4. {2, 4, 6, 8, 9}
5. {2, 5}

6. {1, 2, 4, 7}
7. {3, 5, 9}
8. {1, 2, 3, ..., 10}
9. {2, 4, 6, 8, 10}
10. {1, 3, 5, 7, 9}

EXERCISE 1.8.2

Determine the set of counting numbers described by each of the graphs in Figure 1.3.

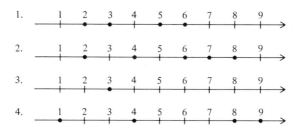

FIGURE 1.3

The set operations of union and intersection can be accomplished by using graphs. For example, to find the union of {3, 5, 6, 8} and {2, 5, 6, 8} each set is graphed onto the same half-line. In Figure 1.4, the elements of

FIGURE 1.4

{3, 5, 6, 8} have been shown by a small ✕ and the elements of {2, 5, 6, 8} have been shown by a small circle. Since the union of {3, 5, 6, 8} and {2, 5, 6, 8} consists of all those elements which belong to either set, or both sets, the graph may be used to determine all those numbers which have been marked by the ✕ or the ◯ or both. Hence,

$$\{3, 5, 6, 8\} \cup \{2, 5, 6, 8\} = \{2, 3, 5, 6, 8\}.$$

A similar procedure may be used to graphically show the intersection of two sets. In Figure 1.5, the elements of {1, 3, 5, 6, 9, 11} have been marked by a small ✕ and the elements of {2, 3, 6, 9} have been marked by a small circle. Since the intersection of {1, 3, 5, 6, 9, 11} and {2, 3, 6, 9} consists of all those elements which belong to both sets, the graph may be used to determine all those elements which have been marked by both the ✕ and the ◯. Hence, {1, 3, 5, 6, 9, 11} ∩ {2, 3, 6, 9} = {3, 6, 9}.

FIGURE 1.5

EXERCISE 1.8.3

Do each of the following problems by graphing the two sets on one half-line and reading the result wanted from the graph.

1. {1, 2, 5, 8, 10} ∪ {3, 4, 5, 8} =
2. {4, 7, 9, 11} ∪ {2, 4, 9, 10, 11} =
3. {1, 2, 3, 4, 5} ∪ {3, 4, 5, 6, 7, 8} =
4. {3, 6, 9, 12} ∩ {2, 4, 6, 8, 10, 12} =
5. {1, 4, 8, 10} ∩ {2, 5, 9, 11} =
6. {4, 5, 6, 7, 8} ∩ {1, 2, 3, 4, 5} =
7. {1, 2, 3, 4} ∪ {4, 5, 6} =
8. {2, 4, 6, 8, 10} ∪ {1, 3, 5, 7, 9} =
9. {1, 2, 3, . . . , 10} ∩ {4, 5, 6} =
10. {1, 2, 3} ∩ {5, 8, 9, 10} =

The graphing of ordered pairs is accomplished by using two half-lines which are perpendicular to each other as shown in Figure 1.6. On the graph the point marked by the letter H is called the origin. To graph the ordered pair (4, 3), the first component, 4, is used to move four units to the right of the origin, and the second component, 3, is used to move three units up. The graph of the ordered pair (4, 3) is shown in Figure 1.7. In Figure 1.7,

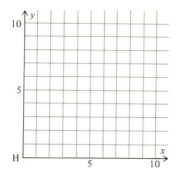

FIGURE 1.6

there is a point marked A. This point designates the ordered pair (9, 2), because it is nine units to the right of the origin and two units up.

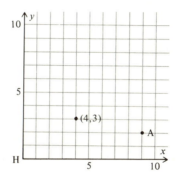

FIGURE 1.7

EXERCISE 1.8.4

Draw a graph for ordered pairs of counting numbers and show the location of each of the following:

|       |         |     |          |
|-------|---------|-----|----------|
| 1.    | (3, 2)  | 6.  | (10, 4)  |
| 2.    | (8, 1)  | 7.  | (6, 2)   |
| 3.    | (1, 5)  | 8.  | (2, 6)   |
| 4.    | (6, 6)  | 9.  | (4, 4)   |
| 5.    | (5, 9)  | 10. | (8, 1)   |

EXERCISE 1.8.5

Figure 1.8 is a graph of ordered pairs. Find the ordered pair for each point, as has been done for point A.

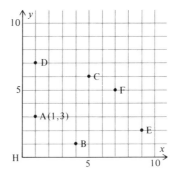

FIGURE 1.8

Sets of ordered pairs may now be graphed by plotting each of the elements of the set. The set $\{(5, 2), (4, 3), (3, 4), (2, 5), (1, 6)\}$ is shown by the graph in figure 1.9.

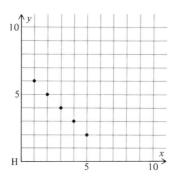

FIGURE 1.9

EXERCISE 1.8.6

Graph each of the following sets of ordered pairs.
1.  $\{(1, 7), (2, 1), (2, 5), (4, 7)\}$
2.  $\{(3, 1), (3, 2), (3, 3), (3, 4), (3, 5), (3, 6), (3, 7)\}$
3.  $\{(1, 2), (2, 1), (3, 2), (4, 3), (5, 4), (6, 5)\}$
4.  $\{(1, 4), (2, 4), (3, 4), (5, 4), (6, 4), (7, 4)\}$
5.  $\{(3, 5), (1, 7), (2, 2), (8, 2)\}$
6.  $\{(1, 2), (2, 3), (3, 4), (4, 5), (5, 6), (6, 7)\}$
7.  $\{(1, 4), (2, 3), (3, 2), (4, 1), (5, 2), (6, 3), (7, 4)\}$
8.  $\{(4, 1), (5, 2), (6, 3), (7, 4), (6, 5), (5, 6), (4, 7),$
    $(3, 6), (2, 5), (1, 4), (2, 3), (3, 2)\}$

EXERCISE 1.8.7

Find the Cartesian Product of the following sets and plot the graphs.

1.  $\{3, 5, 6\} \times \{1, 4, 5\}$     5.  $\{2, 4, 7\} \times \{1, 3, 5\}$
2.  $\{1, 3\} \times \{2, 5, 6, 7\}$     6.  $\{1, 6\} \times \{4, 6, 7, 8\}$
3.  $\{5\} \times \{2\}$     7.  $\{8\} \times \{3\}$
4.  $\{1, 2, 3, 4, 5, 6\} \times \{5\}$     8.  $\{2, 4, 6, 8, 10\} \times \{4\}$

# 2 THE COUNTING NUMBERS

## 2.1 Introduction to Set of Counting Numbers

In this chapter we are concerned with the counting numbers, which are designated by the set, $\{1, 2, 3, \ldots\}$. The laws which govern the counting numbers, and the use of these laws in developing other relationships are discussed in this chapter.

The number 1 is the least element in the set, $\{1, 2, 3, \ldots\}$, but the set of counting numbers has no greatest element. No matter how high we count we could still count higher if necessary. If one were to suggest that the set had a greatest element, then it would only be necessary to count one more to show there is yet another element which is greater. The three dots in the set, $\{1, 2, 3, \ldots\}$, indicate that the counting numbers have no greatest element.

$\{1, 2, 3, \ldots\}$ is called the set of counting numbers because its elements are used to count objects, such as the number of desks in the room, the number of days in a week, and the number of dollars in your savings account.

Though we say we count the number of days in a week we actually make a pairing of the days in the week with elements of the set of counting numbers. This can be shown by the following:

$$
\begin{array}{lcl}
\text{Monday} & \leftrightarrow & 1 \\
\text{Tuesday} & \leftrightarrow & 2 \\
\text{Wednesday} & \leftrightarrow & 3 \\
\text{Thursday} & \leftrightarrow & 4 \\
\text{Friday} & \leftrightarrow & 5 \\
\text{Saturday} & \leftrightarrow & 6 \\
\text{Sunday} & \leftrightarrow & 7 \\
\end{array}
$$

FIGURE 2.1

In Figure 2.1 each day of the week is paired with a counting number. Hence, there must be the same number of days in a week as there are elements in the set, $\{1, 2, 3, 4, 5, 6, 7\}$. Since 7 is the greatest number in $\{1, 2, 3, 4, 5, 6, 7\}$ we say there are seven days in a week.

Another example of the counting process would be to find the number of elements in the set, $\{\sqrt{}, +, <, @, =\}$. The elements of the set can be paired with the counting numbers as shown below:

$$
\begin{array}{lcl}
\sqrt{} & \leftrightarrow & 1 \\
+ & \leftrightarrow & 2 \\
< & \leftrightarrow & 3 \\
@ & \leftrightarrow & 4 \\
= & \leftrightarrow & 5 \\
\end{array}
$$

FIGURE 2.2

Since each element of $\{\sqrt{}, +, <, @, =\}$ has been paired with an element of $\{1, 2, 3, 4, 5\}$ and 5 is the highest number or greatest element in the set, we say that $\{\sqrt{}, +, <, @, =\}$ has five elements.

Perhaps the two preceding examples of counting the days in a week and the number of elements in $\{\sqrt{}, +, <, @, =\}$ seem to make the counting procedure unnecessarily difficult. While it is rarely true that we make a written list of the numbers we use to count the elements of a set, we invariably make such a relationship mentally. For example, find the number of elements in the set, $\{7, 12, 19, 36, 42, 81, 96, 104, 201\}$. To answer this problem you have probably pointed to each of the elements in the set, and mentally assigned the numbers $\{1, 2, 3, 4, 5, 6, 7, 8, 9\}$ to the elements. In fact, you have made, mentally, the following pairings:

$$
\begin{array}{ccc}
7 & \leftrightarrow & 1 \\
12 & \leftrightarrow & 2 \\
19 & \leftrightarrow & 3 \\
36 & \leftrightarrow & 4 \\
42 & \leftrightarrow & 5 \\
81 & \leftrightarrow & 6 \\
96 & \leftrightarrow & 7 \\
104 & \leftrightarrow & 8 \\
201 & \leftrightarrow & 9
\end{array}
$$

FIGURE 2.3

Hence we say {7, 12, 19, 36, 42, 81, 96, 104, 201} has nine elements because its elements have been paired with {1, 2, 3, 4, 5, 6, 7, 8, 9}.

When two sets can be associated so that each element of the first set is paired exactly with one element of the second set, and each element of the second set is paired exactly with one element of the first set, we say the sets are in one-to-one correspondence.

The set of days in the week can be paired in a one-to-one correspondence with {1, 2, 3, 4, 5, 6, 7}. {√, +, <, @, =} can be paired in one-to-one correspondence with {1, 2, 3, 4, 5}. {7, 12, 19, 36, 42, 81, 96, 104, 201} can be paired in a one-to-one correspondence with {1, 2, 3, ..., 9}.

EXERCISE 2.1.1

Write a set of counting numbers that can be placed in a one-to-one correspondence with each given set, and find the number of elements in each set.

1. {×, ÷, <}
2. The set of months in a year.
3. {4, 8, 9, 10, 11, 17, 19, 31}
4. {I, II, III, IV, V, ..., XIII}
5. The set of states in the United States.
6. The set of letters in the English alphabet.
7. {5, 13, 51, 73, 81, 406, 603}
8. {6, 7, 8, ..., 20}
9. The set of continents.
10. {2, 4, 6, ..., 76}

## 2.2  Addition of Counting Numbers

The addition of counting numbers is directly related to the counting process. To add $2 + 3$ is to find the total number of elements in $\boxed{\#\quad\#}$ and $\boxed{\#\quad\#\quad\#}$. This is comparable to counting the elements in $\boxed{\#\quad\#\quad\#\quad\#\quad\#}$.

Another way of describing addition in terms of counting relies upon the union of sets and a one-to-one correspondence between sets. $\{1, 2\}$ is a set containing two numerals. $\{I, II, III\}$ is a set containing three numerals. The union of these sets of numerals is $\{1, 2, I, II, III\}$, which can be paired in a one-to-one correspondence with $\{1, 2, 3, 4, 5\}$, and therefore has five elements.

The addition of $4 + 5$ can be explained in terms of counting as follows:

$\{1, 2, 3, 4\}$ has four numerals as elements.
$\{I, II, III, IV, V\}$ has five numerals as elements.
$\{1, 2, 3, 4\} \cup \{I, II, III, IV, V\} = \{1, 2, 3, 4, I, II, III, IV, V\}$
$\{1, 2, 3, 4, I, II, III, IV, V\}$ can be placed in a one-to-one correspondence with $\{1, 2, 3, 4, 5, 6, 7, 8, 9\}$. Therefore, a set of four numerals and a set of five numerals gives us a set of nine numerals, and $4 + 5 = 9$.

EXERCISE 2.2.1

1.  (a) How many numerals are elements in $\{1, 2, 3, 4\}$?
    (b) How many numerals are elements in $\{I, II\}$?
    (c) How many numerals are elements in $\{1, 2, 3, 4\} \cup \{I, II\}$?
    (d) Since $\{1, 2, 3, 4, I, II\}$ can be placed in a one-to-one correspondence with $\{1, 2, 3, 4, 5, 6\}$, $4 + 2$ is ____.
2.  (a) How many numerals are elements in $\{1, 2, 3, \ldots, 11\}$?
    (b) How many numerals are elements in $\{I, II, III, IV, V\}$?
    (c) Find a set of counting numbers that can be placed in a one-to-one correspondence with
        $\{1, 2, 3, \ldots, 11\} \cup \{I, II, III, IV, V\}$.
    (d) How many elements are there in
        $\{1, 2, 3, \ldots, 11\} \cup \{I, II, III, IV, V\}$?
    (e) $11 + 5$ is ____.
3.  (a) How many numerals are elements in $\{1, 2, 3, \ldots, 25\}$?
    (b) How many numerals are elements in $\{I, II, III, \ldots, XII\}$?
    (c) How many numerals are elements in
        $\{1, 2, 3, \ldots, 25\} \cup \{I, II, III, \ldots, XII\}$?

(d) Find a set of counting numbers that can be placed in a one-to-one correspondence with
$\{1, 2, 3, \ldots, 25\} \cup \{\text{I, II, III}, \ldots, \text{XII}\}$.

(e) How many elements are there in
$\{1, 2, 3, \ldots, 25\} \cup \{\text{I, II, III}, \ldots, \text{XII}\}$?

(f) $25 + 12$ is ____.

4. (a) How many elements are in the set, $\{1, 2, 3, \ldots, 251\}$?

(b) How many elements are in the set, $\{\text{I, II, III}, \ldots, \text{XXVII}\}$?

(c) How many elements are in the set,
$\{1, 2, 3, \ldots, 251\} \cup \{\text{I, II, III}, \ldots, \text{XXVII}\}$?

(d) Find a set of counting numbers that can be placed in a one-to-one correspondence with
$\{1, 2, 3, \ldots, 251\} \cup \{\text{I, II, III}, \ldots, \text{XXVII}\}$.

(e) $251 + 27$ is ____.

The procedure used in the preceding exercise is ample evidence of the claim that the addition of counting numbers is completely dependent upon the ability to count. Although it is assumed that you have already acquired a proficiency at addition of counting numbers, the viewpoint of addition as counting will be helpful in understanding the remainder of this chapter.

One direct benefit that can be derived from the addition procedure of the last exercise is a recognition that addition can be performed only on two numbers at a time. The fact follows directly from the use of the set operation, union, which was used in the exercise. Since union can only be accomplished on two sets at a time, the addition of counting numbers is also restricted to only two numbers at a time.

$(5 + 3) + 7$ is a numerical expression. The plus symbols $(+)$ indicate that the operation of addition is to be performed. Since there are three numbers in the expression, $(5 + 3) + 7$, and only two numbers can be added at one time, the parentheses are used to show that the operation addition is to be performed on 5 and 3 first. To evaluate the numerical expression, $(5 + 3) + 7$, means to find the counting number named by the expression. This is accomplished in two steps as follows:

$$(5 + 3) + 7$$

$$8 + 7$$

$$15$$

The numerical expression, $[7 + (4 + 9)] + 5$, involves both parentheses and square brackets. Parentheses ( ) and square brackets [ ] have the same purpose in mathematics. Both are used so there will be no question regarding the grouping of numbers in an expression. There are three numbers in the square brackets and two in the parentheses. Hence, the evaluation of $[7 + (4 + 9)] + 5$ is accomplished in the following steps:

$$[7 + (4 + 9)] + 5$$

$$[7 + 13] + 5$$

$$20 + 5$$

$$25$$

(Notice that the operation addition was used on only two counting numbers to perform each step.)

In the numerical expression $(9 + 6) + [8 + 1]$, there are both parentheses and square brackets. Since each grouping symbol contains two numbers the evaluation of $(9 + 6) + [8 + 1]$ is accomplished by two steps as follows:

$$(9 + 6) + [8 + 1]$$

$$15 + 9$$

$$24$$

EXERCISE 2.2.2

1.  Complete *only the first step* in evaluating each of the following numerical expressions.
    (a) $4 + (9 + 5)$            (d) $[9 + 7] + 4$
    (b) $[6 + 3] + (5 + 2)$      (e) $(6 + [3 + 1]) + 8$
    (c) $4 + [8 + (7 + 22)]$     (f) $(4 + 11) + [8 + 19]$
2.  Evaluate each numerical expression completely.
    (a) $(8 + 3) + 9$            (c) $(7 + 7) + [6 + 5]$
    (b) $[9 + (3 + 2)] + 4$      (d) $6 + (8 + [3 + 2])$
3.  Do $17 + 23$ and $23 + 17$ have the same evaluation?
4.  Do $9 + (8 + 4)$ and $(9 + 8) + 4$ have the same evaluation?
5.  Do $[6 + 3] + 13$ and $6 + [3 + 13]$ have the same evaluation?
6.  Do $(5 + 4) + (3 + 7)$ and $5 + [(4 + 3) + 7]$ have the same evaluation?
7.  Do $41 + 18$ and $18 + 41$ have the same evaluation?
8.  Do $603 + 94$ and $94 + 603$ have the same evaluation?
9.  Do $(7 + 8) + 13$ and $7 + (8 + 13)$ have the same evaluation?
10. Do $[6 + (3 + 7)] + 2$ and $(6 + 3) + [7 + 2]$ have the same evaluation?

$7 + 19$ is a numerical expression which names the counting number 26. 26 is called the sum of 7 and 19, because the evaluation of $7 + 19$ is 26. Two important properties of addition are illustrated by the example of

7 + 19 = 26. First, there is one and only one correct sum for any addition expression involving counting numbers. That is, the answer for any addition problem is *unique*. Secondly, the sum of two counting numbers is always itself a counting number. For example, 9 + 47 = 56 and 104 + 257 = 361, where the sums, 56 and 361, are themselves counting numbers. The fact that the sum of two counting numbers is always a counting number is called the *Closure Property for Addition* for the set of counting numbers.

## 2.3  Subtraction of Counting Numbers

The properties mentioned in the preceding paragraph may not seem surprising, but they are not shared by all operations that are commonly used with counting numbers. Subtraction is an example of such an operation.

13 − 7 is a numerical expression which names the unique counting number 6. The operation subtraction can be completely explained in terms of addition. In fact, the evaluation of any subtraction expression may be accomplished by answering an addition question.

The subtraction expression 19 − 8, can be evaluated by correctly replacing the question mark in 8 + ? = 19.

47 − 32 can be answered by replacing the expression by 32 + ? = 47. Every subtraction expression can be evaluated correctly by answering an addition question.

EXERCISE 2.3.1

Write each of the following subtraction problems as an addition question.

1.  12 − 5 = ?      6.  6 − 11 = ?
2.  92 − 46 = ?     7.  15 − 9 = ?
3.  205 − 19 = ?    8.  63 − 15 = ?
4.  257 − 102 = ?   9.  47 − 905 = ?
5.  14 − 10 = ?    10.  55 − 10 = ?

6 − 11 = ? can be written as 11 + ? = 6. However, there is no counting number that can be added to 11 to give a sum of 6. Therefore, 6 − 11 = ? is a subtraction expression which has no counting number evaluation. Comparing subtraction with addition, we find that subtraction does not have the two important properties enjoyed by addition. First, not every subtraction expression names one and only one counting number, because some subtraction expressions do not have a counting number evaluation. Secondly, there is no closure property for subtraction of counting numbers because the evaluation of a subtraction expression is not necessarily a counting number.

EXERCISE 2.3.2

Each of the following problems is a numerical expression. If the evaluation is a counting number, find it; if not, state there is no counting number evaluation.

| | | | |
|---|---|---|---|
| 1. | $17 - 3$ | 7. | $84 + 14$ |
| 2. | $3 - 17$ | 8. | $14 - 84$ |
| 3. | $17 + 3$ | 9. | $63 + 18$ |
| 4. | $3 + 17$ | 10. | $18 + 63$ |
| 5. | $84 - 14$ | 11. | $18 - 63$ |
| 6. | $14 + 84$ | 12. | $63 - 18$ |

Like addition, subtraction is an operation performed on only two numbers at a time. The expression, $15 - [6 - (8 - 5)]$, involves four counting numbers and is evaluated as follows:

$$15 - [6 - (8 - 5)]$$
$$15 - [6 - 3]$$
$$15 - 3$$
$$12$$

The expression $(12 - 5) - [(17 - 9) - 1]$ has no counting number evaluation as illustrated in the following steps:

$$(12 - 5) - [(17 - 9) - 1]$$
$$7 - [8 - 1]$$
$$7 - 7$$

Since zero is not a counting number, $7 - 7$, has no counting number evaluation.

EXERCISE 2.3.3

1. Evaluate or state that the expression has no counting number evaluation.
   (a) $11 - (15 - 10)$      (d) $4 - (15 - 13)$
   (b) $(17 - 5) - 8$      (e) $12 - [(4 - 1) - 2]$
   (c) $(14 - 11) - 7$      (f) $[6 - (8 - 5)] - 4$
2. Do $815 + 47$ and $47 + 815$ have the same evaluation?
3. Do $27 - 16$ and $16 - 27$ have the same evaluation?
4. Do $16 + (9 + 4)$ and $(16 + 9) + 4$ have the same evaluation?

5. Do $(19 - 6) - 4$ and $19 - (6 - 4)$ have the same evaluation?
6. Addition of counting numbers_____have the closure property.
   (does, does not)
7. Subtraction of counting numbers_____have the closure property.
   (does, does not)

## 2.4 Multiplication of Counting Numbers

The addition of counting numbers can be completely explained in terms of counting by using the concepts of union and one-to-one correspondence of sets. Recall that $\{1, 2, 3, \ldots, 11\}$ has eleven numerals and $\{I, II, III, \ldots, XXII\}$ has twenty two numerals as elements. The union of these two sets of numerals can be placed in a one-to-one correspondence with $\{1, 2, 3, \ldots, 33\}$. Hence, $11 + 22$ has a sum of 33.

The multiplication of counting numbers can also be described completely in terms of counting. $\{1, 2, 3, 4\}$ has four elements. $\{1, 2, 3\}$ has three elements. $\{1, 2, 3, 4\} \times \{1, 2, 3\}$ is the set of ordered pairs

$$\left\{ \begin{array}{l} (1, 1), (1, 2), (1, 3), (2, 1), (2, 2), (2, 3), \\ (3, 1), (3, 2), (3, 3), (4, 1), (4, 2), (4, 3) \end{array} \right\}$$

Since the above set of ordered pairs can be placed in a one-to-one correspondence with $\{1, 2, 3, \ldots, 12\}$, the multiplication of four and three names the counting number 12.

EXERCISE 2.4.1

1. How many elements are in the set, $\{1, 2\}$?
2. How many elements are in the set, $\{1, 2, 3\}$?
3. Find the Cartesian Product, $\{1, 2\} \times \{1, 2, 3\}$.
4. How many elements (ordered pairs) are in $\{1, 2\} \times \{1, 2, 3\}$?
5. $2 \times 3 =$ _____.
6. How many elements are in $\{6, 7\}$?
7. How many elements are in $\{2, 3, 4, 5\}$?
8. Find the Cartesian Product of $\{6, 7\} \times \{2, 3, 4, 5\}$.
9. How many elements are in $\{6, 7\} \times \{2, 3, 4, 5\}$?
10. $2 \times 4 =$ _____.
11. How many elements are in $\{13\}$?
12. How many elements are in $\{1, 2, 3, \ldots, 7\}$?
13. Find the Cartesian Product of $\{13\} \times \{1, 2, 3, \ldots, 7\}$.
14. How many elements are in $\{13\} \times \{1, 2, 3, \ldots, 7\}$?
15. $1 \times 7 =$ _____.

16. In finding the Cartesian Product of $\{1, 2, 3, 4\}$ and $\{11, 12, 13, 14, 15\}$, how many times would each element of the first set, $\{1, 2, 3, 4\}$, appear as a first component of an ordered pair?

17. Find the number of ordered pairs in each of the following:
   (a) $\{1, 2, 3, \ldots, 9\} \times \{1, 2\}$
   (b) $\{1, 2, 3\} \times \{1, 2, 3, \ldots, 7\}$
   (c) $\{2, 5, 7, 9\} \times \{8, 9, 10, 11, 19, 26\}$
   (d) $\{1, 2, 3, \ldots, 11\} \times \{1, 2, 3, \ldots, 6\}$
   (e) $\{1, 2, 3, \ldots, 73\} \times \{1, 2, 3, \ldots, 20\}$

18. Evaluate:
   (a) $9 \times 2$                         (d) $11 \times 6$
   (b) $3 \times 7$                         (e) $73 \times 20$
   (c) $4 \times 6$

The symbol for multiplication commonly used in algebra is the dot $(\cdot)$, and shall be used in place of the arithmetic symbol $(\times)$ in the remainder of this text.

$5 \cdot 3$ is a numerical expression. The dot indicates that the operation multiplication is to be performed using the factors 5 and 3. The number 15 is called the product of $5 \cdot 3$, because the evaluation of the multiplication expression is its product.

$(4 \cdot 6) \cdot 2$ is a multiplication expression involving three counting numbers. The parentheses are used to show that $(4 \cdot 6)$ is to be evaluated first. Grouping symbols are necessary because only two counting numbers can be multiplied at one time.

The evaluation of $[4 \cdot (5 \cdot 2)] \cdot 9$ is accomplished in the following steps:

$$[4 \cdot (5 \cdot 2)] \cdot 9$$

$$[4 \cdot 10] \cdot 9$$

$$40 \cdot 9$$

$$360$$

Notice that only two counting numbers were evaluated in each step when performing the operation multiplication.

EXERCISE 2.4.2

1. Complete the first step only in evaluating each of the following numerical expressions:
   (a) $4 \cdot (9 \cdot 5)$                     (d) $(7 \cdot 6) \cdot 13$
   (b) $[6 \cdot 3] \cdot (4 \cdot 5)$              (e) $[8 \cdot (9 \cdot 7)] \cdot 2$
   (c) $6 \cdot [(8 \cdot 4) \cdot 3]$              (f) $(4 \cdot 7) \cdot [8 \cdot 7]$

2.  Evaluate completely:
    (a)  $(5 \cdot 3) \cdot 4$             (c)  $(5 \cdot 5) \cdot [2 \cdot 4]$
    (b)  $[8 \cdot (3 \cdot 2)] \cdot 5$       (d)  $9 \cdot [3 \cdot (1 \cdot 2)]$
3.  Do $9 \cdot 13$ and $13 \cdot 9$ have the same evaluation?
4.  Do $(4 \cdot 7) \cdot 3$ and $4 \cdot (7 \cdot 3)$ have the same evaluation?
5.  Do $(8 \cdot 5) \cdot 4$ and $8 \cdot (5 \cdot 4)$ have the same evaluation?
6.  Do $(4 \cdot 3) \cdot (8 \cdot 5)$ and $[4 \cdot (3 \cdot 8)] \cdot 5$ have the same evaluation?
7.  Do $7 \cdot 31$ and $31 \cdot 7$ have the same evaluation?
8.  Do $61 \cdot 506$ and $506 \cdot 61$ have the same evaluation?
9.  Do $(6 \cdot 10) \cdot 9$ and $6 \cdot (10 \cdot 9)$ have the same evaluation?
10.  Do $(4 \cdot 3) \cdot [7 \cdot 2]$ and $4 \cdot [(3 \cdot 7) \cdot 2]$ have the same evaluation?

## 2.5  Division of Counting Numbers

$35 \div 7$ is a numerical expression which names the unique counting number 5. 5 is called the quotient of 35 and 7, because the evaluation of $35 \div 7$ is 5.

Just as every subtraction expression can be evaluated in terms of an addition question, it is possible to evaluate every division expression in terms of a multiplication question.

The division expression, $42 \div 6$, can be evaluated correctly by replacing the question mark in $6 \cdot ? = 42$.

$54 \div 6$ can be evaluated by correctly replacing the question mark in $6 \cdot ? = 54$.

EXERCISE 2.5.1

Every division expression can be evaluated by answering a multiplication question. Replace each of the following division problems with a multiplication question.

1.  $40 \div 8 = ?$       6.  $56 \div 8 = ?$
2.  $21 \div 7 = ?$       7.  $63 \div 7 = ?$
3.  $240 \div 15 = ?$     8.  $16 \div 20 = ?$
4.  $18 \div 7 = ?$       9.  $8041 \div 11 = ?$
5.  $61 \div 3 = ?$      10.  $59 \div 8 = ?$

$31 \div 5 = ?$ can be replaced with $5 \cdot ? = 31$. There is no counting number that can be multiplied by 5 to give 31. Therefore, $31 \div 5$ has no counting number evaluation.

This situation is very similar to one found earlier with regard to subtraction of counting numbers. The subtraction of counting numbers does not necessarily result in a counting number. Similarly, the division of two counting numbers may or may not result in a counting number. Division of counting numbers does not have the closure property.

EXERCISE 2.5.2

Each of the following is a numerical expression. If the evaluation is a counting number find it; if not, state that there is no counting number evaluation.

1.  $14 \div 7$               6.  $30 \cdot 6$
2.  $7 \div 14$               7.  $6 \div 30$
3.  $7 \cdot 14$              8.  $6 \cdot 30$
4.  $14 \cdot 7$              9.  $56 \div 8$
5.  $30 \div 6$              10.  $8 \div 56$

Division expressions involving three or more counting numbers require parentheses or square brackets to indicate which two counting numbers are to be divided.

The numerical expression $12 \div [(24 \div 3) \div 2]$ is evaluated in the following example:

$$12 \div [(24 \div 3) \div 2]$$

$$12 \div [8 \div 2]$$

$$12 \div 4$$

$$3$$

The expression $[48 \div 6] \div ([63 \div 7] \div 3)$ has no counting number evaluation as shown by the following example:

$$[48 \div 6] \div ([63 \div 7] \div 3)$$

$$8 \div (9 \div 3)$$

$$8 \div 3$$

Since there is no counting number that can be multiplied by 3 to give 8 there is no counting number evaluation for $8 \div 3$.

EXERCISE 2.5.3

1.  Evaluate or state that the expression has no counting number evaluation.
    (a)  $16 \div (44 \div 11)$               (d)  $20 \div (40 \div 8)$
    (b)  $(42 \div 2) \div 7$                 (e)  $60 \div [(100 \div 10) \div 5]$
    (c)  $(18 \div 9) \div 4$                 (f)  $[48 \div (56 \div 7)] \div 4$
2.  Do $937 \cdot 48$ and $48 \cdot 937$ have the same evaluation?
3.  Do $27 \div 9$ and $9 \div 27$ have the same evaluation?

4.  Do $(19 \cdot 6) \cdot 53$ and $19 \cdot (6 \cdot 53)$ have the same evaluation?
5.  Do $(12 \div 6) \div 2$ and $12 \div (6 \div 2)$ have the same evaluation?
6.  Multiplication of counting numbers_____have the closure property.
    <u>(does, does not)</u>
7.  Division of counting numbers_____have the closure property.
    <u>(does, does not)</u>

## 2.6  Numerical Expressions Involving both Addition and Multiplication

$7 + (3 \cdot 6)$, $(4 + 3) \cdot 7$, and $(8 \cdot 2) + [(5 + 2) \cdot 3]$ are numerical expressions which require both multiplication and addition to complete their evaluations. In each case the grouping symbols (parentheses or square brackets) indicate the order in which the operations are to be performed.

$7 + (3 \cdot 6)$ is evaluated in the following steps:

$$7 + (3 \cdot 6)$$

$$7 + 18$$

$$25$$

$(4 + 3) \cdot 7$ is evaluated in the following steps:

$$(4 + 3) \cdot 7$$

$$7 \cdot 7$$

$$49$$

$(8 \cdot 2) + [(5 + 2) \cdot 3]$ is evaluated in the following steps:

$$(8 \cdot 2) + [(5 + 2) \cdot 3]$$

$$16 + [7 \cdot 3]$$

$$16 + 21$$

$$37$$

EXERCISE 2.6.1

Evaluate each of the following numerical expressions.

1.  $(7 \cdot 6) + 5$
2.  $[8 + 3] \cdot 6$
3.  $4 + (8 \cdot 2)$
4.  $5 \cdot (2 + 6)$
5.  $[4 \cdot 3] + (5 + 2)$
6.  $(9 + 3) \cdot [4 + 1]$
7.  $8 + [(5 \cdot 3) + 2]$
8.  $[(9 + 2) \cdot 3] + 6$
9.  $[(8 \cdot 3) + 5] + 4$
10.  $2 + [(8 + 1) \cdot 3]$

5 + 3·4 is a numerical expression involving three counting numbers, but no parentheses are present to indicate whether the first step in evaluating the expression is the addition of 5 + 3 or the multiplication of 3·4. If 5 + 3·4 is evaluated as (5 + 3)·4, the result is 8·4 or 32. If 5 + 3·4 is evaluated as 5 + (3·4), the result is 5 + 12 or 17. Since (5 + 3)·4 = 32 and 5 + (3·4) = 17, it is necessary to have an agreement for evaluating 5 + 3·4 so that only one correct answer will be obtained. This is the intent of the following rule of procedure.

> Whenever a numerical expression involves both multiplication and addition without grouping symbols to indicate the order in which the operations are to be performed, the multiplication operation is performed first.

This rule of procedure takes effect only when no grouping symbols are present to indicate the order of the operations. In such cases it requires that the multiplication be accomplished before the addition, as shown in the following examples:

$$\text{(a)} \qquad 3 \cdot 7 + 6$$
$$21 + 6$$
$$27$$

$$\text{(b)} \qquad 4 + 7 \cdot 8$$
$$4 + 56$$
$$60$$

$$\text{(c)} \quad 5 + [8 + 3 \cdot 6]$$
$$5 + [8 + 18]$$
$$5 + 26$$
$$31$$

EXERCISE 2.6.2

Evaluate:

1. $5 + 8 \cdot 3$
2. $4 \cdot 7 + 3$
3. $(8 \cdot 2 + 1) + 4 \cdot 3$
4. $8 \cdot 3 + 2 \cdot 6$
5. $(4 + 7) \cdot 3 + 5 \cdot 2$

6. $8 + (3 + 2 \cdot 7)$
7. $[2 \cdot 5 + 1] \cdot (2 + 4 \cdot 3)$
8. $5 \cdot (4 + 3) + 2 \cdot 7$
9. $4 \cdot (1 + 6 \cdot 2)$
10. $8 + (3 + 2 \cdot 5)$

## 2.7   Evaluating Open Expressions

$4 + 3 \cdot (7 + 6)$ and $9 + 6 \cdot 4$ are numerical expressions. Each names a unique counting number when evaluated.

$\square + 6$ is an open expression. It is not a numerical expression because the $\square$ is not a counting number. The box is a symbol and represents a position to be replaced by a counting number, but there is no specific counting number that must replace the box. In fact, the box may be replaced by any counting number.

$4 + 5 \cdot \square$ is another open expression. Any counting number may replace the box, but whenever the box is replaced by a counting number a numerical expression will be obtained.

For the open expression, $4 + 5 \cdot \square$,

if $\square$ is replaced by 6, then $4 + 5 \cdot 6$ is obtained,
if $\square$ is replaced by 13, then $4 + 5 \cdot 13$ is obtained, and
if $\square$ is replaced by 104, then $4 + 5 \cdot 104$ is obtained.

For the open expression, $\square \cdot (3 + \square)$,

if $\square$ is replaced by 9, then $9 \cdot (3 + 9)$ is obtained,
if $\square$ is replaced by 20, then $20 \cdot (3 + 20)$ is obtained, and
if $\square$ is replaced by 912, then $912 \cdot (3 + 912)$ is obtained.

Each time the box in an open expression is replaced by a counting number a numerical expression is obtained and the numerical expression may be evaluated.

EXERCISE 2.7.1

1. Evaluate the following open expressions if the box is replaced by 4.
   (a) $5 + \square$              (c) $8 \cdot (\square + 5)$
   (b) $6 \cdot \square + 3$      (d) $\square \cdot (4 + \square)$
2. Evaluate $\square \cdot (8 + \square)$, if the box is replaced by 7.
3. Evaluate $4 \cdot \square + 3$, if the box is replaced by 12.
4. Evaluate $3 \cdot \square + 2$ when the box is replaced by 6
5. Evaluate $8 \cdot \square + 2 \cdot \square$ when the box is replaced by 3
6. Evaluate $4 + 7 \cdot \square$ when the box is replaced by 5
7. Evaluate $\square \cdot (5 + \square)$ when the box is replaced by 2
8. Evaluate $9 + 3 \cdot \square$ when the box is replaced by 8
9. Evaluate $6 \cdot \square + 7 \cdot \square$ when the box is replaced by 2
10. Evaluate $8 \cdot \square + \square \cdot \square$ when the box is replaced by 6

$5\cdot\square + 3$ is an open expression in which the box may be replaced by any counting number. In algebra, letters of the alphabet such as $x$, $y$, and $z$ are used instead of boxes to designate positions that may be replaced by numbers. These letters are called variables.

$5\cdot x + 3$ is an open expression in which the $x$ may be replaced by any counting number. $2 + 4\cdot y$ is another open expression in which the $y$ may be replaced by any counting number.

The multiplication dot in $4\cdot z$ or $6\cdot y$ is unnecessary. Whenever a number is followed by a letter without an operation sign between them, the operation multiplication is understood. $4\cdot z$ may be written as $4z$. $6\cdot y$ may be written $6y$. When replacing $z$ by 6 in the open expression $4z$, the numerical expression must include the multiplication dot, and be written as $4\cdot 6$.

In open expressions such as $6y$, $4z$, and $17x$, the numbers 6, 4, and 17 are called numerical *coefficients* of the variables.

EXERCISE 2.7.2

Evaluate each open expression for the counting number replacement indicated.

1. $x + 5$, when $x = 7$.
2. $3x$, when $x = 8$.
3. $4x + 2$, when $x = 5$.
4. $9 + (y + 6)$, when $y = 3$.
5. $8\cdot(y + 3)$, when $y = 2$.
6. $5y + 4y$, when $y = 6$.
7. $8 + 5z$, when $z = 4$.
8. $(z + 9) + 11$, when $z = 10$.
9. $5w + 4$, when $w = 8$.
10. $4\cdot(w + 6)$, when $w = 7$.
11. $x\cdot(5 + x)$, when $x = 4$.
12. $8w + 3w$, when $w = 5$.
13. Do $5x + 2$ and $7x$ have the same evaluation when $x = 3$?
14. Do $3y + 8$ and $11y$ have the same evaluation when $y = 6$?
15. Do $4x$ and $x + 3x$ have the same evaluation when $x = 9$?
16. Do $4y + 7y$ and $11y$ have the same evaluation when $y = 5$?
17. Do $4\cdot(x + 3)$ and $4x + 3$ have the same evaluation when $x = 7$?
18. Do $5\cdot(y + 2)$ and $5y + 10$ have the same evaluation when $y = 6$?
19. Do $5x + 3$ and $8x$ have the same evaluation when $x = 9$?
20. Do $4(x + 6)$ and $4x + 6$ have the same evaluation when $x = 2$?

In the open expression $5a + b \cdot (4 + c)$ there are three letters to be replaced by counting numbers. Each of the letters, $a$, $b$, and $c$ is called a variable. Each variable may be replaced by any counting number.

To evaluate $5a + b \cdot (4 + c)$ when $a = 6$, $b = 8$, and $c = 2$, the following steps are used:

$$5a + b \cdot (4 + c)$$

$$5 \cdot 6 + 8 \cdot (4 + 2)$$

$$30 + 8 \cdot 6$$

$$30 + 48$$

$$78$$

EXERCISE 2.7.3

1.  Evaluate $3x + y$, when $x = 4$ and $y = 7$.
2.  Evaluate $x + 8y$, when $x = 3$ and $y = 2$.
3.  Evaluate $(4 + x) + 3w$, when $x = 8$ and $w = 4$.
4.  Evaluate $y \cdot (3a + 5)$, when $y = 2$ and $a = 6$.
5.  Evaluate $(3r + 5s) + 2t$, when $r = 2$, $s = 6$, and $t = 4$.
6.  Evaluate $a + [5b + (c + 6)]$, when $a = 4, b = 1$, and $c = 7$.
7.  Do $x + y$ and $y + x$ have the same evaluation when $x = 4$ and $y = 7$?
8.  Do $3x + 2y$ and $5xy$ have the same evaluation when $x = 4$ and $y = 3$?
9.  Do $(x + 2) + y$ and $(x + y) + 2$ have the same evaluation when $x = 5$ and $y = 10$?
10. Do $3x + (8y + 2x)$ and $5x + 8y$ have the same evaluation when $x = 3$ and $y = 2$?

## 2.8   Mathematical Statements

$4 \cdot 3 + 7 = 3 + 2 \cdot 8$ is a mathematical statement which consists of two numerical expressions, $4 \cdot 3 + 7$ and $3 + 2 \cdot 8$, which are connected by an equal sign. The statement claims that the two numerical expressions have the same evaluation. Since both $4 \cdot 3 + 7$ and $3 + 2 \cdot 8$ have an evaluation of 19, the mathematical statement is true.

$8 \cdot 3 + 6 = 4 \cdot 7 + 5$ is a mathematical statement which claims that $8 \cdot 3 + 6$ has the same evaluation as $4 \cdot 7 + 5$. Since $8 \cdot 3 + 6 = 30$ and $4 \cdot 7 + 5 = 33$, the statement is false.

As observed in the previous two paragraphs, mathematical statements can be either true or false.

EXERCISE 2.8.1

Determine whether each of the following mathematical statements is true or false.

1. $4 \cdot 9 + 1 = 37$
2. $(2 + 5) \cdot 8 = 5 \cdot 10 + 6$
3. $8 \cdot 3 + 2 \cdot 3 = 10 \cdot 3$
4. $4 \cdot 5 + 2 = 6 \cdot 5$
5. $2 \cdot (8 + 5) = 2 \cdot 8 + 5$
6. $4 \cdot 7 + 4 \cdot 3 = 4 \cdot (7 + 3)$
7. $9 \cdot 6 + 4 = 13 \cdot 6$
8. $10 + 15 = 5 \cdot (2 + 3)$
9. $4 + 7 = 7 + 4$
10. $(9 + 7) + 13 = 9 + (7 + 13)$

## 2.9  Open Sentences

$2x + 6 = 3x + 1$ is an open sentence which is neither true nor false. The variable, $x$, represents a position to be replaced by a counting number. When $x$ is replaced by a counting number a mathematical statement which may be either true or false is obtained.

If $x = 8$, the open sentence $2x + 6 = 3x + 1$, becomes

$$2 \cdot 8 + 6 = 3 \cdot 8 + 1,$$

which is false.

If $x = 5$, the open sentence $2x + 6 = 3x + 1$ becomes the mathematical statement $2 \cdot 5 + 6 = 3 \cdot 5 + 1$, which is true.

$3a + 2b = 7a + b$ is an open sentence with two variables, $a$ and $b$. $3a + 2b = 7a + b$ is neither true nor false. It will become either a true or false statement whenever the variables, $a$ and $b$, are replaced by counting numbers.

If $a = 3$ and $b = 5$, then $3a + 2b = 7a + b$ becomes

$$3 \cdot 3 + 2 \cdot 5 = 7 \cdot 3 + 5,$$

which is a false mathematical statement.

If $a = 1$ and $b = 4$, then $3a + 2b = 7a + b$ becomes

$$3 \cdot 1 + 2 \cdot 4 = 7 \cdot 1 + 4,$$

which is a true mathematical statement.

EXERCISE 2.9.1

For each of the following open sentences, three numbers are given as replacement values. Determine whether the mathematical statements obtained from these replacements are true or false.

1.  $3a + 7 = 5 \cdot (a + 1)$
    (a)  $a = 1$
    (b)  $a = 3$
    (c)  $a = 6$
2.  $7x + 6 = 13x$
    (a)  $x = 2$
    (b)  $x = 4$
    (c)  $x = 5$
3.  $y + 7 = 7 + y$
    (a)  $y = 4$
    (b)  $y = 6$
    (c)  $y = 10$
4.  $5 \cdot (w + 2) = 4w + 16$
    (a)  $w = 2$
    (b)  $w = 4$
    (c)  $w = 6$
5.  $4x + 5y = x + 8y$
    (a)  $x = 4$   and   $y = 2$
    (b)  $x = 2$   and   $y = 2$
    (c)  $x = 1$   and   $y = 1$
6.  $(3x + 2y) + 5x = 8x + 2y$
    (a)  $x = 1$   and   $y = 3$
    (b)  $x = 1$   and   $y = 4$
    (c)  $x = 3$   and   $y = 2$
7.  $(x + 9) + 13 = (13 + x) + 9$
    (a)  $x = 2$
    (b)  $x = 4$
    (c)  $x = 7$
8.  $5 \cdot (x + 6) = 5x + 30$
    (a)  $x = 2$
    (b)  $x = 4$
    (c)  $x = 9$
9.  $4x + 7 = 11x$
    (a)  $x = 3$
    (b)  $x = 5$
    (c)  $x = 1$

10.  $5x + 9x = 14x$
   (a)  $x = 2$
   (b)  $x = 5$
   (c)  $x = 10$

## 2.10   The Commutative and Associative Laws of Addition

Some open sentences will result in true statements regardless of the numbers used to replace the variables. Two such open sentences are

$$x + y = y + x$$

and

$$(x + y) + z = x + (y + z)$$

For the open sentence $x + y = y + x$, the open expression $x + y$ is equivalent to $y + x$, because any replacements of the variables, $x$ and $y$, will result in a true statement for $x + y = y + x$. This fact is known as the Commutative Law of Addition.

*The Commutative Law of Addition*

> The open expression $x + y$ is *equivalent* to $y + x$. The sum of two counting numbers is the same regardless of the *order* in which they are added.

For the open sentence $(x + y) + z = x + (y + z)$, the open expression $(x + y) + z$ is equivalent to $x + (y + z)$, because any replacements of the variables $x$, $y$, and $z$, will result in a true statement for

$$(x + y) + z = x + (y + z)$$

This is known as the Associative Law of Addition.

*The Associative Law of Addition*

> The open expression $(x + y) + z$ is *equivalent* to $x + (y + z)$. The sum of three counting numbers is the same regardless of the *grouping* obtained by the placement of parentheses.

The Commutative and Associative Laws of Addition may be used to simplify many open expressions involving addition, because the combination of these laws allows any addition expression to be reordered and regrouped.

For example, $(8 + 3x) + 5$ can be simplified by using the following steps:

$(8 + 3x) + 5$

$(3x + 8) + 5$   The Commutative Law allows a change in order.

$3x + (8 + 5)$   The Associative Law allows a change in grouping.

$3x + 13$         Arithmetic fact, $8 + 5 = 13$.

Hence, $(8 + 3x) + 5$ is equivalent to $3x + 13$, which means that any replacement for the variable, $x$, in $(8 + 3x) + 5$ or $3x + 13$ will result in the same evaluation.

Notice that $3x + 13$ *can not* be simplified to $16x$. This is because $3x + 13$ and $16x$ are *not* equivalent. If $x = 5$, then $3x + 13$ has an evaluation of 28, and $16x$ has an evaluation of 80.

EXERCISE 2.10.1

1.   State the property (Commutative or Associative) shown in each of the following open sentences:
   (a)  $7 + x = x + 7$
   (b)  $(4 + y) + 9 = 4 + (y + 9)$
   (c)  $3x + 5y = 5y + 3x$
   (d)  $(9x + 3y) + 2y = 9x + (3y + 2y)$
   (e)  $(6 + x) + 9 = (x + 6) + 9$
   (f)  $5x + (2 + 3x) = 5x + (3x + 2)$
2.   Simplify each of the following open expressions:
   (a)  $(10 + a) + 7$
   (b)  $6 + (r + 8)$
   (c)  $(x + 8) + 5$
   (d)  $(4 + x) + 11$
   (e)  $(y + 16) + 1$
   (f)  $(6 + w) + 13$
   (g)  $14 + (3 + 5x)$
   (h)  $(7 + 9y) + 2$

## 2.11   The Commutative and Associative Laws of Multiplication

$xy = yx$ and $(xy)z = x(yz)$ are two open sentences which become true statements for any replacements of the variables. These open sentences are similar to those used as the basis for the Commutative and Associative Laws of Addition. As such, they are the basis for the following two laws for multiplication of counting numbers.

*The Commutative Law of Multiplication*

The open expression $xy$ is equivalent to $yx$ because any replacements of the variables $x$ and $y$ in $xy = yx$ will result in a true numerical statement. The product of two counting numbers is the same regardless of the *order* in which they are multiplied.

*The Associative Law of Multiplication*

The open expression $(xy)z$ is equivalent to $x(yz)$ because any replacements of the variables $x$, $y$, and $z$ will result in a true mathematical statement. The product of three counting numbers is the same regardless of the *grouping* obtained by the placement of the parentheses.

These two laws of multiplication can be used to simplify many open expressions involving the operation multiplication. The Commutative Law of Multiplication allows the order of two numbers in multiplication to be changed. The Associative Law of Multiplication allows the grouping of three numbers to be changed.

For example, $(6x) \cdot 3$ can be simplified as follows:

$(6x) \cdot 3$

$3 \cdot (6x)$     The Commutative Law allows the order to be changed.

$(3 \cdot 6) \cdot x$     The Associative Law allows the grouping to be changed.

$18x$     Arithmetic fact, $3 \cdot 6 = 18$.

EXERCISE 2.11.1

1.  State the property (Commutative or Associative) shown in each of the following open sentences.
    (a) $5 \cdot (3x) = (5 \cdot 3) \cdot x$
    (b) $7x = x \cdot 7$
    (c) $(7x) \cdot 4 = 4 \cdot (7x)$
    (d) $(4x) \cdot y = 4 \cdot (xy)$
    (e) $8 \cdot (wz) = 8 \cdot (zw)$
    (f) $(3a) \cdot 5 = 5 \cdot (3a)$

2. Find a simpler equivalent expression for each of the following:
   (a) $6 \cdot (5x)$
   (b) $(8w) \cdot 3$
   (c) $5 \cdot (4z)$
   (d) $(2x) \cdot 9$
   (e) $(4a) \cdot 7$
   (f) $6 \cdot (2b)$

## 2.12 The Multiplication Law of One and the Distributive Law of Multiplication over Addition

There are two more laws for counting numbers which must be considered. The first of these is so obvious that its usefulness is often overlooked.

$1x = x$ is an open sentence which will become a true statement for any replacement for the variable $x$. The fact that $1x$ is equivalent to $x$ is due to a multiplication property of 1, which states when one is multiplied by a counting number the product is the same as the counting number.

Some examples of the fact that one multiplied by any counting number gives a product of that same counting number are as follows:

$$1 \cdot 7 = 7$$

$$1 \cdot 59 = 59$$

$$1 \cdot (6 \cdot 3 + 2) = 6 \cdot 3 + 2$$

*The Multiplication Law of One*

The open expressions $1x$ and $x$ are equivalent because 1 multiplied by any counting number gives that same counting number.

Some examples of the Multiplication Law of One applied to open expressions are as follows:

$$1x = x$$

$$7z + z = 7z + 1z$$

$$1 \cdot (3x + 2) = 3x + 2$$

The last law for the set of counting numbers is the only law which concerns both multiplication and addition. In the numerical expression $4 \cdot (5 + 2)$, the parentheses indicate that $5 + 2$ is to be evaluated first, and the sum then multiplied by 4. The evaluation of $4 \cdot (5 + 2)$ is $4 \cdot 7$ or 28.

In the numerical expression, $4 \cdot 5 + 4 \cdot 2$, the lack of parentheses indicates that $4 \cdot 5$ and $4 \cdot 2$ are to be evaluated first. The evaluation is then completed by taking the sum of the two products. The evaluation of $4 \cdot 5 + 4 \cdot 2$ is $20 + 8$ or $28$.

Notice that $4 \cdot (5 + 2)$ and $4 \cdot 5 + 4 \cdot 2$ have the same evaluation. Whenever a number is multiplied by a sum the same result can be found by evaluating the sum of the products. Some examples of this property are as follows:

$$6 \cdot (3 + 4) = \quad 6 \cdot 7 \quad = 42$$

$$6 \cdot 3 + 6 \cdot 4 = 18 + 24 = 42$$

$$5 \cdot (6 + 2) = \quad 5 \cdot 8 \quad = 40$$

$$5 \cdot 6 + 5 \cdot 2 = 30 + 10 = 40$$

*The Distributive Law of Multiplication over Addition*

$x \cdot (y + z)$ is equivalent to $xy + xz$, because $x \cdot (y + z) = xy + xz$ becomes a true statement for any counting number replacements for the variables, $x$, $y$, and $z$.

The Distributive Law of Multiplication over Addition is used to simplify open expressions as follows:

According to the Distributive Law, $5 \cdot (x + 4)$ is equivalent to $5 \cdot x + 5 \cdot 4$. To simplify $5 \cdot (x + 4)$, the following steps are used:

$$5 \cdot (x + 4)$$

$$5 \cdot x + 5 \cdot 4$$

$$5x + 20$$

To simplify $7 \cdot (3 + x)$, the following steps are used:

$$7 \cdot (3 + x)$$

$$7 \cdot 3 + 7 \cdot x$$

$$21 + 7x$$

According to the Distributive Law, $5x + 7x$ is equivalent to $(5 + 7) \cdot x$. To simplify $5x + 7x$, the following steps are used:

$$5x + 7x$$

$$(5 + 7) \cdot x$$

$$12x$$

To simplify $8x + 3x$, the following steps are used:

$$8x + 3x$$

$$(8 + 3) \cdot x$$

$$11x$$

To simplify $7x + x$, the Multiplication Law of One and the Distributive Property are used as follows:

$7x + x$

$7x + 1x$     Multiplication Law of One

$(7 + 1) \cdot x$     Distributive Law

$8x$     Arithmetic fact, $7 + 1 = 8$.

EXERCISE 2.12.1

1.  State the property (Multiplication Law of One or the Distributive Law of Multiplication over Addition) that is shown in each of the following open sentences.
    (a) $1w = w$
    (b) $5 \cdot (4 + x) = 5 \cdot 4 + 5 \cdot x$
    (c) $1 \cdot (2x + 19) = 2x + 19$
    (d) $4x + 11x = (4 + 11) \cdot x$
    (e) $1 \cdot (17y) = 17y$
    (f) $6 \cdot (x + 3) = 6x + 6 \cdot 3$
2.  Simplify each of the following:
    (a) $4 \cdot (x + 3)$
    (b) $8 \cdot (2 + y)$
    (c) $2x + 3x$
    (d) $1 \cdot (9x + 2)$
    (e) $3 \cdot (x + 9)$
    (f) $1 \cdot (9z)$
    (g) $5x + x$
    (h) $2 \cdot (x + 8)$

The eight laws of the counting numbers have been discussed in the preceding sections. These laws with their equivalent expressions are as follows:

The Closure Law of Addition

The Closure Law of Multiplication

| | |
|---|---|
| The Commutative Law of Addition | $x + y = y + x$ |
| The Associative Law of Addition | $(x + y) + z = x + (y + z)$ |
| The Commutative Law of Multiplication | $xy = yx$ |
| The Associative Law of Multiplication | $(xy)z = x(yz)$ |
| The Multiplication Law of One | $1x = x$ |
| The Distributive Law of Multiplication over Addition | $x(y + z) = xy + xz$ |

The simplification of open expressions is an important algebraic skill, but such a skill is directly traceable to an understanding of the eight laws for the counting numbers. Every simplification is accomplished through the application of one or more of these eight laws and the basic arithmetic combinations. The following examples are to explain the application of these laws in the simplification of a variety of open expressions.

(a)  $5 + 6x + 3$

| | |
|---|---|
| $(6x + 5) + 3$ | Commutative Law of Addition |
| $6x + (5 + 3)$ | Associative Law of Addition |
| $6x + 8$ | Arithmetic fact $5 + 3 = 8$. |

(b)  $3 + 9x + 3x$

| | |
|---|---|
| $3 + (9x + 3x)$ | Associative Law of Addition |
| $3 + (9 + 3) \cdot x$ | Distributive Law of Multiplication over Addition |
| $3 + 12x$ | Arithmetic fact $9 + 3 = 12$. |

(c)  $6 + 3 \cdot (x + 7)$

| | |
|---|---|
| $6 + 3 \cdot x + 3 \cdot 7$ | Distributive Law of Multiplication over Addition |
| $6 + 3x + 21$ | Arithmetic fact, $3 \cdot 7 = 21$. |
| $(3x + 6) + 21$ | Commutative Law of Addition |
| $3x + (6 + 21)$ | Associative Law of Addition |
| $3x + 27$ | Arithmetic fact, $6 + 21 = 27$. |

(d)  $4 \cdot (3x + 5)$

$\quad\quad 4 \cdot (3x) + 4 \cdot 5$      Distributive Law of Multiplication over Addition

$\quad\quad (4 \cdot 3) \cdot x + 4 \cdot 5$      Associative Law of Multiplication

$\quad\quad 12x + 20$      Arithmetic facts, $4 \cdot 3 = 12$ and $4 \cdot 5 = 20$.

EXERCISE 2.12.2

Simplify:

1.  $9 + (x + 4)$
2.  $(z + 6) + 3$
3.  $5 \cdot (4y)$
4.  $(2x) \cdot 8$
5.  $5x + 2x$
6.  $3x + x$
7.  $5 \cdot (x + 4)$
8.  $3 \cdot (5 + y)$
9.  $4 \cdot (5x + 1)$
10.  $8 \cdot (2x + 3)$
11.  $4x + 3 + 5 + 2x$
12.  $8x + 7 + 9x + 1$
13.  $2y + 4x + 7x + y$
14.  $7 + x + 4x + 8x$
15.  $2 + 3 \cdot (x + 6)$
16.  $5 \cdot (2 + y) + 2y$
17.  $4 \cdot (2x + 1) + 5 \cdot (x + 3)$
18.  $9 + 3 \cdot (x + 7) + x$
19.  $4x + 5 \cdot (2 + 3x) + 2x$
20.  $9 \cdot (x + 5) + 2 \cdot (3x + 7)$

## 2.13  Solving Simple Equations

$x + 5 = 7$ is an open sentence in which $x$ may be replaced by any counting number. Some replacements for $x$ will result in false statements. For example, if $x = 4$, then $x + 5 = 7$ becomes the numerical statement, $4 + 5 = 7$, which is false.

There is one replacement for $x$ that will make $x + 5 = 7$ a true statement. If $x = 2$, then $x + 5 = 7$ becomes the numerical statement, $2 + 5 = 7$, which is true.

$3x = 21$ is another open sentence in which $x$ can be replaced by any counting number. If $x$ is replaced by 5, then $3x = 21$ becomes the numerical statement $3 \cdot 5 = 21$, which is false. If $x$ is replaced by 7, then $3x = 21$ becomes the true statement, $3 \cdot 7 = 21$.

Open sentences such as $x + 5 = 7$ and $3x = 21$ are often called equations. To solve an equation is to find the replacement(s) for the variable that will result in a true statement.

To solve $x + 8 = 23$ ask the question, "What counting number can be added to 8 to give a sum of 23?"

To solve $7x = 56$ ask the question, "What counting number can be multiplied by 7 to give a product of 56?"

Other methods for solving equations will be shown as the equations become more difficult. For the time being, however, the student will find that no other method is really necessary.

EXERCISE 2.13.1

Solve each equation by finding a replacement for the variable that will result in a true statement.

| | | | |
|---|---|---|---|
| 1. | $x + 7 = 11$ | 11. | $4x = 28$ |
| 2. | $4 + x = 9$ | 12. | $5x = 15$ |
| 3. | $21 = 6 + x$ | 13. | $3x = 12$ |
| 4. | $14 + x = 27$ | 14. | $42 = 6x$ |
| 5. | $19 = x + 1$ | 15. | $16 = 2x$ |
| 6. | $x + 43 = 50$ | 16. | $9x = 81$ |
| 7. | $400 + x = 435$ | 17. | $6x = 6$ |
| 8. | $28 = 19 + x$ | 18. | $30x = 90$ |
| 9. | $47 + x = 100$ | 19. | $180 = 10x$ |
| 10. | $25 = x + 21$ | 20. | $460 = 2x$ |

## 2.14  Solving Equations Requiring Simplification

To solve $(8 + x) + 3 = 16$, the left side of the equation, $(8 + x) + 3$, should first be simplified to $x + 11$. Hence, $(8 + x) + 3 = 16$ and $x + 11 = 16$ will have the same solution.

$$(8 + x) + 3 = 16$$

$$x + 11 = 16$$

5 is the solution

To solve $4x + 5x = 27$, the left side of the equation, $4x + 5x$, should first be simplified to $9x$. Hence, $4x + 5x = 27$ and $9x = 27$ will have the same solution.

$$4x + 5x = 27$$

$$9x = 27$$

3 is the solution

To solve $4 \cdot (3x) = 24$, the left side of the equation, $4 \cdot (3x)$, should first be simplified to $12x$. Hence $4 \cdot (3x) = 24$ and $12x = 24$ will have the same solution.

$$4 \cdot (3x) = 24$$

$$12x = 24$$

2 is the solution

EXERCISE 2.14.1

Solve each equation by first simplifying the open expression involving the variable.

1. $(x + 3) + 5 = 12$
2. $(4 + x) + 11 = 23$
3. $37 = 19 + (x + 1)$
4. $16 + (4 + y) = 53$
5. $(9 + y) + 6 = 30$
6. $14 = (y + 6) + 3$
7. $(5 + y) + 14 = 93$
8. $6x + 4x = 40$
9. $3x + 5x = 24$
10. $8x + x = 36$

11. $2x + 5x = 63$
12. $14 = x + x$
13. $9 = 2x + x$
14. $3x + 3x = 66$
15. $6 \cdot (2x) = 36$
16. $2 \cdot (4x) = 48$
17. $(2x) \cdot 9 = 18$
18. $100 = 5 \cdot (2x)$
19. $42 = (3x) \cdot 7$
20. $3 \cdot (5x) = 45$

## 2.15   The Domain of the Variable

Every equation is solved in terms of a particular set of elements for replacements of the variable. Such a set is the domain of the variable.

For example, $\{2, 7, 9, 13\}$ can be given as the domain of $x$ for the equation, $2x + 3 = 21$. Using the elements of $\{2, 7, 9, 13\}$ as replacements for $x$ in the equation, $2x + 3 = 21$, the following statements are obtained:

if   $x = 2$,   $2 \cdot 2 + 3 = 21$,   which is false,

if   $x = 7$,   $2 \cdot 7 + 3 = 21$,   which is false,

if   $x = 9$,   $2 \cdot 9 + 3 = 21$,   which is true, and

if   $x = 13$,   $2 \cdot 13 + 3 = 21$,   which is false.

For the domain $\{2, 7, 9, 13\}$ and the equation $2x + 3 = 21$, the singleton set $\{9\}$ is called the *truth set*, because 9 is the only replacement from the domain, that provides a true statement.

Consider another example. Let the domain be $\{1, 5, 7, 12\}$ and the equation be $2 \cdot (x + 4) = 2x + 8$. Replacing $x$ by each element of $\{1, 5, 7, 12\}$, the following statements are obtained:

if   $x = 1$,   $2 \cdot (1 + 4) = 2 \cdot 1 + 8$,   which is true,

if   $x = 5$,   $2 \cdot (5 + 4) = 2 \cdot 5 + 8$,   which is true,

if   $x = 7$,   $2 \cdot (7 + 4) = 2 \cdot 7 + 8$,   which is true, and

if   $x = 12$,   $2 \cdot (12 + 4) = 2 \cdot 12 + 8$,   which is true.

Hence, $\{1, 5, 7, 12\}$ is the truth set of $2 \cdot (x + 4) = 2x + 8$, because every element of the domain provides a true statement.

One last possibility must be illustrated. For a domain of $\{5, 7, 10\}$ and the equation, $3x + 7 = 31$, if $x$ is replaced by each element of the domain the following statements are obtained:

$$\text{if} \quad x = 5, \quad 3 \cdot 5 + 7 = 31, \quad \text{which is false,}$$

$$\text{if} \quad x = 7, \quad 3 \cdot 7 + 7 = 31, \quad \text{which is false, and}$$

$$\text{if} \quad x = 10, \quad 3 \cdot 10 + 7 = 31, \quad \text{which is false.}$$

For the domain, $\{5, 7, 10\}$, the truth set of $3x + 7 = 31$ is the empty set, $\{\ \}$, because there is no element of the domain that will provide a true statement.

EXERCISE 2.15.1

Each problem below contains a domain set and an equation. Find the truth set.

| | | |
|---|---|---|
| 1. | $\{4, 5, 9, 13\}$, | $x + 7 = 16$ |
| 2. | $\{3, 7, 8, 11\}$, | $x + 3 = 9$ |
| 3. | $\{2, 4, 5, 9, 10\}$, | $(x + 5) + 7 = 12 + x$ |
| 4. | $\{5, 9, 11, 13\}$, | $4x = 52$ |
| 5. | $\{3, 4, 7, 8, 12\}$, | $19x = 76$ |
| 6. | $\{2, 4, 6, 7\}$, | $5x + 9 = 39$ |
| 7. | $\{4, 8, 9\}$, | $3x + 5 = 26$ |
| 8. | $\{1, 4, 7, 11, 20\}$, | $8x + 3 = 59$ |
| 9. | $\{2, 7, 8, 9\}$, | $9x + 13 = 31$ |
| 10. | $\{2, 3, 4, 5, 6, 7\}$, | $6x + 8 = 32$ |
| 11. | $\{1, 2, 3, \ldots, 10\}$, | $3x + 2 = 20$ |
| 12. | $\{1, 2, 3, \ldots, 10\}$, | $7x + 5 = 54$ |
| 13. | $\{4, 5, 8, 10\}$, | $6 \cdot (x + 3) = 6x + 18$ |
| 14. | $\{1, 2, 6, 9, 10\}$, | $3x + 9 = x + 13$ |
| 15. | $\{1, 2, 3, 4, 5\}$, | $5x + 7 = 12x$ |
| 16. | $\{1, 2, 3, 4, 5\}$, | $4 \cdot (x + 2) = 4x + 2$ |
| 17. | $\{1, 2, 3, 4, 5\}$, | $6x + 7x = 13x$ |
| 18. | $\{1, 2, 3, 4, 5\}$, | $(2x + 7) + 3x = (5x + 6) + 1$ |
| 19. | $\{1, 2, 3, 4, 5\}$, | $x + 13 = 4x + 1$ |
| 20. | $\{1, 2, 3, 4, 5\}$, | $7x + 9 = 3x + 29$ |

If the set of counting numbers is used as the domain for the equation, $x + 7 = 19$, the question to be asked, and answered, is "what counting numbers can be added to 7 to obtain a sum of 19?" Since the only counting number to satisfy this condition is 12, the truth set of $x + 7 = 19$ is $\{12\}$.

For the domain, $\{1, 2, 3, \ldots\}$, and the equation, $x + 13 = 5$, the question to be answered is "what counting number can be added to 13 to obtain a sum of 5?" There is no counting number to satisfy this condition. Hence, the truth set is the empty set, { }.

Similarly, using the domain, {1, 2, 3, ...}, $3x = 39$ has a truth set of {13}, and $4x = 7$ has a truth set of { }.

EXERCISE 2.15.2

Using the set of counting numbers, {1, 2, 3, ...}, as the domain, find the truth set for each of the following equations:

| | | | |
|---|---|---|---|
| 1. | $x + 4 = 12$ | 11. | $2x = 11$ |
| 2. | $17 + x = 20$ | 12. | $5x = 105$ |
| 3. | $33 = x + 16$ | 13. | $63 = 21x$ |
| 4. | $x + 850 = 900$ | 14. | $18 = 5x$ |
| 5. | $x + 15 = 11$ | 15. | $200x = 800$ |
| 6. | $97 = 43 + x$ | 16. | $743x = 743$ |
| 7. | $x + 187 = 141$ | 17. | $2x + 1 = 7$ |
| 8. | $x + 8 = 30$ | 18. | $3x + 2 = 17$ |
| 9. | $4x = 28$ | 19. | $5x + 4 = 54$ |
| 10. | $60 = 6x$ | 20. | $4x + 3 = 19$ |

## 2.16   Inequalities

">" is a symbol that is read as "greater than." "<" is a symbol that is read as "less than." $8 > 3$ is true because 8 is greater than 3. $5 < 2$ is false because 5 is not less than 2.

$4x < 17$ is an open sentence called an inequality. Using {1, 2, 3, ...} as the domain, the truth set of $4x < 17$ will consist of those counting numbers which when multiplied by 4 will give a product that is *less than* 17. Hence, the truth set is {1, 2, 3, 4}.

$3x > 19$ is an open sentence or inequality. If the domain is {1, 2, 3, ...}, the truth set of $3x > 19$ will consist of all those counting numbers which when multiplied by 3 will give a product that is *greater than* 19. Hence, the truth set is {7, 8, 9, ...}.

Similarly, using {1, 2, 3, ...} as the domain, the inequality $x + 7 > 26$ has a truth set of {20, 21, 22, ...} and the inequality $x + 4 < 81$ has a truth set of {1, 2, 3, ..., 76}.

EXERCISE 2.16.1

Use the domain, {1, 2, 3, ...}, to find the truth set for each of the following inequalities:

| | | | |
|---|---|---|---|
| 1. | $x + 9 > 16$ | 3. | $x + 47 < 91$ |
| 2. | $x + 14 < 37$ | 4. | $x + 23 > 76$ |

5.  $x + 3 > 2$
6.  $x + 17 < 5$
7.  $5x > 31$
8.  $6x > 100$
9.  $7x < 96$
10. $4x < 50$
11. $19x < 6$
12. $12x > 7$

13. $2x + 3 > 14$
14. $3x + 5 < 28$
15. $7x + 6 > 93$
16. $x + 8 > 20$
17. $x + 15 > 52$
18. $x + 5 < 48$
19. $5x < 48$
20. $8x > 71$

## 2.17  Equations with Two Variables

$xy = 12$ is an equation in two variables. If the domain of the variables is $\{1, 2, 3, \ldots\}$, then any counting number can replace $x$, and any counting number can replace $y$. Since it takes a pair of replacements to change $xy = 12$ to a mathematical statement, the solutions will be shown as ordered pairs of counting numbers. $(3, 4)$ is a solution of $xy = 12$, because $3 \cdot 4 = 12$ is a true mathematical statement. $(5, 7)$ is not a solution of $xy = 12$, because if $x = 5$ and $y = 7$, the statement obtained, $5 \cdot 7 = 12$, is false. The set of all ordered pairs $(x, y)$ that would be the truth set of $xy = 12$ for a domain, $\{1, 2, 3, \ldots\}$, is

$$\{(1, 12), (2, 6), (3, 4), (4, 3), (6, 2), (12, 1)\}.$$

The graph of the truth set of $xy = 12$, using ordered pairs, $(x, y)$, where $x$ and $y$ are counting numbers is shown in Figure 2.4.

The equation $2x + y = 11$ is also an equation in two variables. Using $\{1, 2, 3, \ldots\}$ as the domain for the variables, the set of ordered pairs, $(x, y)$,

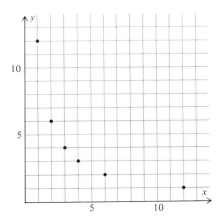

FIGURE 2.4

that would be the truth set of $2x + y = 11$ is

$$\{(1, 9), (2, 7), (3, 5), (4, 3), (5, 1)\}.$$

The graph of this truth set is shown in Figure 2.5.

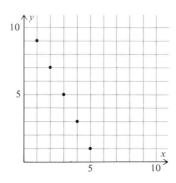

FIGURE 2.5

EXERCISE 2.17.1

Use the set of counting numbers, $\{1, 2, 3, \ldots\}$, as the domain of the variables $x$ and $y$. Find the set of ordered pairs $(x, y)$ that is the truth set of each of the following equations, and graph the truth sets.

| | | | |
|---|---|---|---|
| 1. | $xy = 15$ | 11. | $xy = 18$ |
| 2. | $x + y = 12$ | 12. | $x + y = 10$ |
| 3. | $xy = 20$ | 13. | $xy = 30$ |
| 4. | $x + y = 7$ | 14. | $x + y = 14$ |
| 5. | $2x + y = 15$ | 15. | $2x + y = 9$ |
| 6. | $3x + y = 13$ | 16. | $3x + y = 16$ |
| 7. | $5x + y = 18$ | 17. | $4x + y = 20$ |
| 8. | $6x + y = 9$ | 18. | $5x + y = 26$ |
| 9. | $4x + 3y = 31$ | 19. | $2x + 3y = 18$ |
| 10. | $5x + 2y = 36$ | 20. | $3x + 5y = 28$ |

# 3 THE INTEGERS

## 3.1 Introduction

In chapter 2 the set of counting numbers, the laws for the system of counting numbers, and some of the applications of these laws for simplifying open expressions and solving equations and inequalities were studied. It was shown that some equations of the form $ax = b$ where $a$ and $b$ are counting numbers had a singleton set as the truth set while others had the empty set as the truth set.

For example, using $\{1, 2, 3, \ldots\}$ as the domain, $5x = 20$, has the truth set, $\{4\}$, while $2x = 7$ has the truth set, $\{\;\}$.

The same situation exists for equations of the form $x + a = b$ where $a$ and $b$ are counting numbers. Using $\{1, 2, 3, \ldots\}$ as the domain, the equation, $x + 18 = 24$, has the truth set, $\{6\}$, while $x + 9 = 3$, has the truth set, $\{\;\}$, because there is no counting number that can be added to 9 to produce a sum of 3.

In this chapter another set of numbers, called the *integers* is to be studied. Later we shall see that if the set of integers is used as the domain, every equation of the form $x + a = b$ will have a singleton set as its truth set.

### 3.2    The Set of Integers

$x + 4 = 4$ has no solution in the set of counting numbers because zero is not a counting number. If $\{0, 1, 2, 3, \ldots\}$ is used as the domain for the equation, $x + 4 = 4$, the truth set would be $\{0\}$, because $0 + 4 = 4$ is a true statement. Note that $\{0\}$ is not the same as $\{\ \}$. The empty set has no elements. $\{0\}$ has one element, the number zero.

Equations such as $x + 1 = 0$, $x + 2 = 0$, and $x + 19 = 0$ have no solution in the set $\{1, 2, 3, \ldots\}$. In order for $x + 1 = 0$ to have a number solution it is necessary to expand our domain set to include a number called the *opposite of 1*. The opposite of 1 is shown by the numeral $^-1$. The opposite of 1, $^-1$, has the important addition property that when added to 1, the sum is zero. Hence, $1 + {}^-1 = 0$ is a true statement.

The *opposite of 2* is shown by the numeral $^-2$. The opposite of 2, $^-2$, has the important property that when added to 2, the sum is zero. Hence, $2 + {}^-2 = 0$ is a true statement.

Every counting number has an opposite. The opposite of 19 is $^-19$. The opposite of 412 is shown by the numeral $^-412$. The set of all opposites for the counting numbers can be shown as $\{^-1, {}^-2, {}^-3, \ldots\}$.

The opposite of $^-7$ is 7, because $^-7 + 7 = 0$ is a true statement. The opposite of $^-41$ is 41, because $^-41 + 41 = 0$.

EXERCISE 3.2.1

1.  The opposite of 5 is _____.
2.  The opposite of 24 is _____.
3.  The opposite of $^-37$ is _____.
4.  The opposite of 533 is _____.
5.  The opposite of $^-412$ is _____.
6.  $^-(^-7)$ is the numeral to designate the opposite of the opposite of 7. What simpler numeral may be used for $^-(^-7)$?
7.  What is the opposite of the opposite of 9?
8.  $^-[^-(^-3)]$ is the opposite of the opposite of $^-3$. What simpler numeral may be used for $^-[^-(^-3)]$?
9.  What is the opposite of the opposite of $^-5$?
10. The opposite of 0 is the number that can be added to 0 to give a sum of zero. What is the opposite of 0?

$\{^-1, {}^-2, {}^-3, \ldots\}$ is the set of all opposites for the set of counting numbers. $\{^-1, {}^-2, {}^-3, \ldots\}$ is called the set of *negative integers*. Every counting number has its opposite in the set of negative integers, $\{^-1, {}^-2, {}^-3, \ldots\}$.

{1, 2, 3, ...} is the set of counting numbers, but is also called the set of *positive integers*. Every negative integer has its opposite in the set of positive integers, {1, 2, 3, ...}.

Zero is its own opposite. Zero is not an element of {1, 2, 3, ...} or {⁻1, ⁻2, ⁻3, ...}. Hence, zero is not a positive or a negative integer. Zero is the only number that is its own opposite.

If the set of positive integers is unioned with the set of negative integers and the set, {0}, the result is a set containing all the positive integers, all the negative integers, and zero. This set is the set of *integers*, and is shown as {..., ⁻2, ⁻1, 0, 1, 2, ...}.

Every positive integer is an element of the set, {..., ⁻2, ⁻1, 0, 1, 2, ...}. Every negative integer is an element of the set, {..., ⁻2, ⁻1, 0, 1, 2, ...}. Zero is also an element of the set, {..., ⁻2, ⁻1, 0, 1, 2, ...}.

Every integer has its opposite in the set of integers, which means that every equation of the form $x + b = 0$, where $b$ is an integer will have a singleton set as its truth set when the set of integers, {..., ⁻2, ⁻1, 0, 1, 2, ...}, is used as the domain.

For example, if {..., ⁻2, ⁻1, 0, 1, 2, ...} is used as the domain for the following equations the resulting truth sets will be found to be singleton sets.

$x + 19 = 0$ has {⁻19} as its truth set.

$x + ⁻12 = 0$ has {12} as its truth set.

EXERCISE 3.2.2

1. Is ⁻7 an element of {1, 2, 3, ...}?
2. Is 16 an element of {⁻1, ⁻2, ⁻3, ...}?
3. Is 19 an element of {..., ⁻2, ⁻1, 0, 1, 2, ...}?
4. Is ⁻73 an element of {..., ⁻2, ⁻1, 0, 1, 2, ...}?
5. Is every integer an element of {..., ⁻2, ⁻1, 0, 1, 2, ...}?
6. Does every integer have its opposite in {..., ⁻2, ⁻1, 0, 1, 2, ...}?
7. Is there any integer that is neither positive nor negative?
8. Is 4 equal to ⁻4?
9. Is ⁻18 equal to 18?
10. Is there an integer that is equal to its opposite?
11. Use {..., ⁻2, ⁻1, 0, 1, 2, ...} as the domain to find the truth set for each of the following equations:
    (a) $x + 14 = 0$      (e) $x + 0 = 0$
    (b) $x + ⁻10 = 0$     (f) $x + ⁻(⁻5) = 0$
    (c) $x + ⁻42 = 0$     (g) $x + ⁻[⁻(⁻9)] = 0$
    (d) $x + 81 = 0$

### 3.3  Addition of Integers

The set of integers is often shown graphically on a number line like the one in Figure 3.1. The arrows at the ends of the figure indicate that both the negative and the positive integers continue forever.

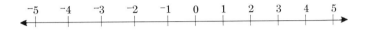

FIGURE 3.1

The use of the graph of figure 3.1 also plainly shows that every integer except zero is not equal to its opposite. This fact must be clearly understood in learning to add integers. Although ⁻5 and 5 are opposites, they are different numbers, and methods for adding opposites and for adding negative numbers must now be developed. Every integer has an opposite. For example, 9 and ⁻9 are opposites and ⁻17 and 17 are opposites.

Before stating any rules for the addition of integers, it is possible to do some problems on the assumption that addition is to be both commutative and associative as was the case with the counting numbers. The commutative law allows the order of numbers in addition to be changed and the associative law allows the grouping of numbers in addition to be changed.

Study the following problems to see the manner in which the commutative and associative laws are used to arrive at the sum. Notice that in each case two opposites are grouped together.

$$^-6 + 4 + 1 + {}^-4 + {}^-1 + 9 + 6$$

$$(^-6 + 6) + (4 + {}^-4) + (1 + {}^-1) + 9$$

$$0 \quad + \quad 0 \quad + \quad 0 \quad + 9$$

$$9$$

$$7 + {}^-5 + {}^-3 + 8 + {}^-7 + 5 + {}^-8$$

$$(7 + {}^-7) + (^-5 + 5) + (8 + {}^-8) + {}^-3$$

$$0 \quad + \quad 0 \quad + \quad 0 \quad + {}^-3$$

$$^-3$$

EXERCISE 3.3.1

Add each of the following problems by grouping pairs of opposites.

1.  $6 + {}^-9 + {}^-6 + 5 + 9 =$
2.  ${}^-4 + {}^-5 + 3 + 4 + {}^-3 =$
3.  $8 + {}^-1 + 5 + {}^-4 + 1 + {}^-8 + {}^-5 =$
4.  ${}^-14 + 12 + {}^-3 + 14 + 3 + {}^-6 + {}^-12 =$
5.  ${}^-9 + 5 + 4 + {}^-6 =$
6.  $4 + {}^-3 + {}^-5 + 3 + 7 + 5 + {}^-6 + 6 =$
7.  ${}^-9 + 5 + {}^-3 + {}^-5 + 2 + 1 =$
8.  $15 + {}^-3 + {}^-7 + 3 + {}^-4 + {}^-15 + 7 =$
9.  $6 + 9 + {}^-6 + 2 + {}^-9 + {}^-2 + 5 =$
10.  ${}^-14 + 8 + {}^-9 + 6 =$

Using the grouping procedures necessary in the last exercise, it is now possible to show why the sum of two negative integers is a negative integer.

The opposite of 10 is ${}^-10$ because ${}^-10 + 10 = 0$. If it can be shown that $({}^-7 + {}^-3) + 10$ also has a sum of zero, then ${}^-7 + {}^-3$ has a sum of ${}^-10$.

$$({}^-7 + {}^-3) + 10$$

$$({}^-7 + {}^-3) + (3 + 7)$$

$$({}^-7 + 7) + ({}^-3 + 3)$$

$$0 + 0$$

$$0$$

Hence, ${}^-7 + {}^-3 = {}^-10$ must be a true statement if addition of integers is to be both commutative and associative.

EXERCISE 3.3.2

1.  Find the sums:
    (a)  $({}^-6 + {}^-9) + 15$          (d)  $({}^-8 + {}^-8) + 16$
    (b)  $({}^-4 + {}^-1) + 5$          (e)  $({}^-3 + {}^-5) + 8$
    (c)  $({}^-12 + {}^-6) + 18$
2.  Find the sums:
    (a)  ${}^-6 + {}^-9$          (d)  ${}^-8 + {}^-8$
    (b)  ${}^-4 + {}^-1$          (e)  ${}^-3 + {}^-5$
    (c)  ${}^-12 + {}^-6$
3.  True or false? The sum of two negative integers is the opposite of the sum of their opposites.

The sum of two negative integers is a negative integer. The sum of two positive integers is a positive integer. These two addition facts are important in understanding the addition of a positive and a negative integer.

To add $5 + {}^-3$, the positive integer 5 is written as $2 + 3$, because 3 is the opposite of $^-3$.

$$5 + {}^-3$$

$$(2 + 3) + {}^-3$$

$$2 + (3 + {}^-3)$$

$$2 + 0$$

$$2$$

Hence, $5 + {}^-3 = 2$ is a true statement because 3 and $^-3$ are opposites and the operation addition is to be associative.

To add $4 + {}^-9$, the negative integer $^-9$ is written as $^-4 + {}^-5$, because $^-4$ is the opposite of 4.

$$4 + {}^-9$$

$$4 + ({}^-4 + {}^-5)$$

$$(4 + {}^-4) + {}^-5$$

$$0 + {}^-5$$

$${}^-5$$

Hence, $4 + {}^-9 = {}^-5$ is a true statement, because $^-4$ and 4 are opposites and addition is to be associative.

Figure 3.2 shows the set of integers graphically. To add $5 + {}^-3$, since 5 is further from zero than $^-3$, we write 5 as $2 + 3$.

FIGURE 3.2

To add $4 + {}^-9$, since $^-9$ is further from zero than 4, we write $^-9$ as $^-4 + {}^-5$.

EXERCISE 3.3.3

1.  Using figure 3.2, determine which integer is further from zero.
    (a) $^-6, 4$           (d) $^-8, 6$
    (b) $^-9, 12$         (e) $^-4, 5$
    (c) $^-2, 1$

2. Complete the following addition problems.
   (a) $^-6 + 4 = (^-2 + ^-4) + 4 =$
   (b) $^-9 + 12 = ^-9 + (9 + 3) =$
   (c) $^-2 + 1 = (^-1 + ^-1) + 1 =$
   (d) $^-8 + 6 = (^-2 + ^-6) + 6 =$
   (e) $^-4 + 5 = ^-4 + (4 + 1) =$
3. Add each of the following:

| | |
|---|---|
| (a) $4 + ^-11$ | (f) $^-10 + 1$ |
| (b) $^-6 + 10$ | (g) $^-8 + 26$ |
| (c) $^-3 + 8$ | (h) $^-27 + 12$ |
| (d) $19 + ^-12$ | (i) $43 + ^-21$ |
| (e) $14 + ^-11$ | (j) $33 + ^-51$ |

## 3.4  Summary for the Addition of Integers

The rules for addition of integers may be summarized as follows:

(a) The sum of any integer and zero is that same integer.

$$13 + 0 = 13, \quad ^-17 + 0 = ^-17, \quad \text{and} \quad 0 + 0 = 0$$

(b) The sum of two positive integers is a positive integer.

$$4 + 5 = 9 \quad \text{and} \quad 19 + 8 = 27$$

(c) The sum of two negative integers is a negative integer.

$$^-9 + ^-5 = ^-14 \quad \text{and} \quad ^-23 + ^-11 = ^-34$$

(d) The sum of a positive and negative integer may be positive, negative, or zero depending upon which integer is further from zero, or in the case of opposites where the distance from zero is the same.

$$13 + ^-7 = 6 \quad \text{and} \quad ^-8 + 23 = 15$$

$$^-11 + 4 = ^-7 \quad \text{and} \quad 5 + ^-17 = ^-12$$

$$^-8 + 8 = 0 \quad \text{and} \quad 47 + ^-47 = 0$$

EXERCISE 3.4.1

| | |
|---|---|
| 1. $5 + ^-7 =$ | 7. $^-8 + ^-3 =$ |
| 2. $6 + 13 =$ | 8. $16 + ^-9 =$ |
| 3. $^-12 + ^-1 =$ | 9. $8 + ^-7 =$ |
| 4. $^-33 + 0 =$ | 10. $14 + ^-20 =$ |
| 5. $^-14 + 10 =$ | 11. $^-19 + ^-7 =$ |
| 6. $9 + ^-3 =$ | 12. $^-7 + 13 =$ |

| | |
|---|---|
| 13.  $12 + 29 =$ | 22.  $(^-4 + ^-8) + 17 =$ |
| 14.  $^-43 + ^-21 =$ | 23.  $18 + (9 + ^-5) =$ |
| 15.  $^-50 + 23 =$ | 24.  $^-11 + (^-6 + ^-1) =$ |
| 16.  $12 + ^-81 =$ | 25.  $7 + (8 + ^-9) =$ |
| 17.  $73 + ^-55 =$ | 26.  $^-10 + (6 + ^-10) =$ |
| 18.  $^-35 + ^-17 =$ | 27.  $(5 + ^-8) + 12 =$ |
| 19.  $0 + 42 =$ | 28.  $(22 + 9) + ^-37 =$ |
| 20.  $61 + ^-98 =$ | 29.  $(^-46 + 15) + 19 =$ |
| 21.  $(6 + ^-3) + ^-5 =$ | 30.  $16 + (^-23 + 2) =$ |

## 3.5   The Minus Sign as an Operation Symbol

The minus sign is commonly associated with the operation subtraction. In chapter 2, there was a discussion of subtraction of counting numbers. It should be recalled that every subtraction problem can be written as an addition question. It should also be recalled that subtraction of counting numbers is neither commutative nor associative, and, therefore, does not have the flexibility enjoyed by addition. In this section the use of the minus sign is completely explained in terms of addition.

$14 - 8 = ?$ can be written as $8 + ? = 14$. Hence, $14 - 8 = \underline{6}$ is a true statement because $8 + \underline{6} = 14$ is a true statement.

$7 - 10 = ?$ can be written as $10 + ? = 7$. Although neither of these problems has a counting number answer, the set of integers offers the solution, $^-3$. $7 - 10 = \underline{^-3}$, because $10 + \underline{^-3} = 7$ is a true statement.

$5 - ^-7 = ?$ can be written as $^-7 + ? = 5$. Consequently, $5 - ^-7 = \underline{12}$, because $^-7 + \underline{12} = 5$ is a true statement.

$^-6 - ^-4 = ?$ can be written as $^-4 + ? = ^-6$. Consequently, $^-6 - ^-4 = \underline{^-2}$, because $^-4 + \underline{^-2} = ^-6$ is a true statement.

EXERCISE 3.5.1

Write each of the following as an addition question, and then find the integer named in the expression.

| | |
|---|---|
| 1.   $4 - 1 = ?$ | 11.  $19 - 12 = ?$ |
| 2.   $12 - 17 = ?$ | 12.  $^-14 - 6 = ?$ |
| 3.   $15 - 6 = ?$ | 13.  $^-11 - 1 = ?$ |
| 4.   $14 - 23 = ?$ | 14.  $14 - ^-3 = ?$ |
| 5.   $9 - ^-6 = ?$ | 15.  $5 - ^-7 = ?$ |
| 6.   $8 - ^-10 = ?$ | 16.  $^-4 - ^-3 = ?$ |
| 7.   $^-6 - 5 = ?$ | 17.  $^-5 - ^-5 = ?$ |
| 8.   $^-2 - 8 = ?$ | 18.  $3 - 8 = ?$ |
| 9.   $^-9 - ^-2 = ?$ | 19.  $12 - 1 = ?$ |
| 10.  $^-3 - ^-7 = ?$ | 20.  $16 - ^-3 = ?$ |

In the previous exercise numerical expressions involving minus signs were evaluated by first writing an addition question. The procedure for evaluating such expressions can now be further simplified.

$5 - 3$ names the same integer as $5 + {}^-3$. Notice that two sign changes were made. Instead of subtracting 3 from 5, the opposite of 3 was added to 5. $5 - 3 = 5 + {}^-3$.

$4 - 9$ names the same integer as $4 + {}^-9$. Again, two sign changes were made. The minus sign was changed to addition, and 9 was changed to its opposite, ${}^-9$. $4 - 9 = 4 + {}^-9$.

$6 - {}^-7$ names the same integer as $6 + 7$. The same two sign changes should now be evident. The expression is written as an addition expression by adding the opposite of ${}^-7$. $6 - {}^-7 = 6 + 7 = 13$.

Every numerical expression involving a minus sign can be written as an addition expression by adding the opposite of the second number. Using variables where ${}^-x$ is read as "the opposite of $x$," this statement can be expressed in general terms as, $y - x = y + {}^-x$. Numerical examples of this general statement are as follows:

$$5 - 9 = 5 + {}^-9$$

$$6 - {}^-10 = 6 + 10$$

$${}^-4 - {}^-3 = {}^-4 + 3$$

$${}^-8 - 2 = {}^-8 + {}^-2$$

EXERCISE 3.5.2

1. Write each of the following minus expressions as addition expressions:
   (a) $6 - 13$
   (b) $7 - {}^-2$
   (c) ${}^-11 - {}^-6$
   (d) ${}^-5 - 13$
2. Evaluate:
   (a) $4 - 12$
   (b) $19 - {}^-7$
   (c) ${}^-21 - {}^-3$
   (d) ${}^-12 - 1$
   (e) $4 - {}^-7$
   (f) ${}^-6 - {}^-1$
   (g) $15 - 4$
   (h) ${}^-16 - {}^-16$
   (i) $13 - {}^-2$
   (j) $4 - 21$

## 3.6 Evaluating Numerical Expressions Involving Addition

$(5 + {}^-7) + {}^-4$ is a numerical expression involving the addition of integers. The parentheses indicate that $5 + {}^-7$ is to be evaluated first.

$$(5 + {}^-7) + {}^-4$$

$${}^-2 + {}^-4$$

$${}^-6$$

$(6 - 7) - {}^-9$ is a numerical expression involving minus signs that should be evaluated as an addition expression:

$$(6 - 7) - {}^-9$$

$$(6 + {}^-7) + 9$$

$${}^-1 + 9$$

$$8$$

${}^-6 - ({}^-8 - 3)$ is a numerical expression with a minus sign immediately preceding the parentheses, and should be evaluated as follows:

$${}^-6 - ({}^-8 - 3)$$

$${}^-6 - ({}^-8 + {}^-3)$$

$${}^-6 - {}^-11$$

$${}^-6 + 11$$

$$5$$

EXERCISE 3.6.1

Evaluate:

1. $(4 - 6) - 9$
2. $5 + ({}^-3 - 2)$
3. $9 - (8 - 7)$
4. $({}^-3 - 11) - {}^-2$
5. ${}^-8 - (6 - 11)$

6. ${}^-3 - ({}^-12 + {}^-1)$
7. $(5 - {}^-3) - (8 + {}^-1)$
8. $({}^-4 - 8) - (9 + {}^-3)$
9. $(10 - 3) + ({}^-8 - {}^-2)$
10. $(13 - 27) - (11 - 6)$

## 3.7 The Multiplication of Integers

The multiplication of positive integers is actually the same operation as the multiplication of counting numbers. $5 \cdot 3$ means $3 + 3 + 3 + 3 + 3$. Therefore, $5 \cdot 3 = 15$.

Visualizing multiplication as repeated addition it is relatively easy to develop the rule for multiplying a positive integer by a negative integer.

$$6 \cdot {}^-2 = {}^-2 + {}^-2 + {}^-2 + {}^-2 + {}^-2 + {}^-2$$

Evaluating the addition expression, we find that $6 \cdot {}^-2$ has an evaluation of ${}^-12$.

$$5 \cdot {}^-8 = {}^-8 + {}^-8 + {}^-8 + {}^-8 + {}^-8$$

Evaluating the addition expression, ${}^-8 + {}^-8 + {}^-8 + {}^-8 + {}^-8$, we find that $5 \cdot {}^-8$ has an evaluation of ${}^-40$.

EXERCISE 3.7.1

Evaluate each of the following multiplication expressions by considering multiplication as repeated addition:

1. $2 \cdot {}^{-}9$
2. $8 \cdot {}^{-}3$
3. $4 \cdot {}^{-}7$
4. $5 \cdot {}^{-}12$
5. $3 \cdot {}^{-}6$
6. $7 \cdot {}^{-}7$
7. $13 \cdot {}^{-}4$
8. $27 \cdot {}^{-}5$
9. $9 \cdot {}^{-}1$
10. $4 \cdot {}^{-}17$

$2 \cdot {}^{-}9$ has an evaluation of $^{-}18$, because $^{-}9 + {}^{-}9 = {}^{-}18$. If multiplication of integers is to be commutative, since $2 \cdot {}^{-}9$ is $^{-}18$, then $^{-}9 \cdot 2$ must also be $^{-}18$.

$4 \cdot {}^{-}5 = {}^{-}20$. The commutative law of multiplication states that the product of two numbers is the same regardless of the order in which they are taken. Therefore, $^{-}5 \cdot 4$ must also be $^{-}20$.

Because a positive times a negative integer is a negative integer, and because multiplication is to be commutative, we can state one rule for the multiplication of integers:

> The product of two integers, when one is positive and the other is negative, is a negative integer.

EXERCISE 3.7.2

Evaluate:

1. $6 \cdot {}^{-}6$
2. $^{-}9 \cdot 3$
3. $^{-}15 \cdot 4$
4. $4 \cdot {}^{-}8$
5. $^{-}7 \cdot 6$
6. $9 \cdot {}^{-}4$
7. $23 \cdot {}^{-}7$
8. $15 \cdot {}^{-}19$
9. $^{-}41 \cdot 22$
10. $819 \cdot {}^{-}7$
11. If multiplication is considered as repeated addition, what is the evaluation of $8 \cdot 0$?
12. If multiplication is considered as repeated addition, what is the evaluation of $0 \cdot {}^{-}7$?

Whenever zero is multiplied by an integer the product is zero. This fact, along with the distributive law of multiplication over addition explains the multiplication of two negative integers.

Recall the distributive law of multiplication over addition as stated for the counting numbers. $x(y + z) = xy + xz$ is a true statement for any replacements of $x$, $y$, and $z$.

If the distributive law is to be true for the integers, then each of the following equalities will be true statements.

$$5(^-4 + 3) = 5 \cdot {^-4} + 5 \cdot 3$$

$$^-6(10 + {^-7}) = {^-6} \cdot 10 + {^-6} \cdot {^-7}$$

$$^-8(3 + {^-3}) = {^-8} \cdot 3 + {^-8} \cdot {^-3}.$$

The last equality is of special importance to us in showing that $^-8 \cdot {^-3}$ has an evaluation of 24.

$$^-8(3 + {^-3}) = {^-8} \cdot 0 = 0$$

and

$$^-8 \cdot 3 + {^-8} \cdot {^-3} = {^-24} + {^-8} \cdot {^-3}$$

Since the first expressions in the above examples are equal, we can conclude that $^-24 + {^-8} \cdot {^-3} = 0$. Hence, $^-8 \cdot {^-3}$ must have an evaluation of 24, because $^-24 + 24 = 0$.

As another example, consider the following:

$$^-4(9 + {^-9}) = {^-4} \cdot 9 + {^-4} \cdot {^-9} = {^-36} + {^-4} \cdot {^-9}$$

and

$$^-4(9 + {^-9}) = {^-4} \cdot 0 = 0$$

Therefore, $^-36 + {^-4} \cdot {^-9} = 0$, and since $^-36 + 36 = 0$, we conclude that $^-4 \cdot {^-9}$ has an evaluation of 36.

The two examples that show $^-8 \cdot {^-3} = 24$ and $^-4 \cdot {^-9} = 36$ are illustrations of the following rule for the multiplication of two negative numbers.

> The product of two negative integers is
> a positive integer.

EXERCISE 3.7.3

Evaluate:

|     |              |     |              |
|-----|--------------|-----|--------------|
| 1.  | $^-6 \cdot {^-7}$   | 6.  | $^-18 \cdot {^-3}$  |
| 2.  | $^-5 \cdot {^-14}$  | 7.  | $^-58 \cdot {^-10}$ |
| 3.  | $^-2 \cdot {^-4}$   | 8.  | $^-319 \cdot {^-1}$ |
| 4.  | $^-9 \cdot {^-8}$   | 9.  | $^-83 \cdot {^-6}$  |
| 5.  | $^-1 \cdot {^-73}$  | 10. | $^-13 \cdot {^-21}$ |

### 3.8   Summary of the Rules for the Multiplication of Integers

(a)   The product of any integer and zero is zero.

$$12 \cdot 0 = 0$$

(b)   The product of two positive integers is a positive integer.

$$8 \cdot 7 = 56$$

(c)   The product of two negative integers is a positive integer.

$$^-5 \cdot {^-8} = 40$$

(d)   The product of a positive integer and a negative integer is a negative integer.

$$3 \cdot {^-7} = {^-21}$$

EXERCISE 3.8.1

Evaluate:

| | |
|---|---|
| 1.  $4 \cdot {^-7}$ | 11.  $^-3 \cdot {^-6}$ |
| 2.  $^-8 \cdot {^-2}$ | 12.  $^-5 \cdot 7$ |
| 3.  $3 \cdot 9$ | 13.  $4 \cdot 11$ |
| 4.  $^-9 \cdot 6$ | 14.  $^-9 \cdot {^-8}$ |
| 5.  $^-10 \cdot {^-28}$ | 15.  $6 \cdot {^-6}$ |
| 6.  $8 \cdot {^-7}$ | 16.  $^-14 \cdot {^-3}$ |
| 7.  $^-3 \cdot 6$ | 17.  $8 \cdot 15$ |
| 8.  $^-14 \cdot 0$ | 18.  $0 \cdot {^-19}$ |
| 9.  $4 \cdot 12$ | 19.  $^-1 \cdot 73$ |
| 10.  $^-9 \cdot {^-7}$ | 20.  $12 \cdot {^-13}$ |

## 3.9  Evaluating Numerical Expressions Involving Multiplication

$(^-6 \cdot 3) \cdot {^-2}$ is a multiplication expression involving three integers. The parentheses in the expression, $(^-6 \cdot 3) \cdot {^-2}$, indicate that the first step in the evaluation of the expression is to multiply $^-6$ and 3.

$$(^-6 \cdot 3) \cdot {^-2}$$

$$^-18 \cdot {^-2}$$

$$36$$

$4 \cdot (^-3 \cdot 10)$ is evaluated as follows:

$$4 \cdot (^-3 \cdot 10)$$

$$4 \cdot {^-30}$$

$$^-120$$

EXERCISE 3.9.1

Evaluate:

1. $(^-6 \cdot 5) \cdot 2$
2. $^-8 \cdot (^-2 \cdot ^-5)$
3. $4 \cdot (^-3 \cdot ^-5)$
4. $(^-5 \cdot 4) \cdot ^-5$
5. $(^-7 \cdot 2) \cdot 3$
6. $^-6 \cdot (^-7 \cdot ^-2)$
7. $(^-2 \cdot 5) \cdot (3 \cdot ^-4)$
8. $[(^-3 \cdot ^-3) \cdot ^-3] \cdot ^-3$
9. $(4 \cdot 4) \cdot 4$
10. $[(10 \cdot 10) \cdot 10] \cdot 10$
11. Do $4 \cdot (^-3 \cdot ^-7)$ and $(4 \cdot ^-3) \cdot ^-7$ have the same evaluation?
12. Do $(^-8 \cdot ^-5) \cdot ^-6$ and $^-8 \cdot (^-5 \cdot ^-6)$ have the same evaluation?
13. Does the placement of parentheses in a multiplication expression change the evaluation?

## 3.10  Exponents

In the multiplication expression, $7 \cdot (^-2 \cdot 3)$, the integers 7, $^-2$, and 3 are called *factors*. Similarly $^-4$, $^-8$, and 5 are factors of $(^-4 \cdot ^-8) \cdot 5$.

The multiplication expression $7 \cdot 7$ has two factors of 7, and can be written as the expression $7^2$. The 2 in the expression, $7^2$ is called an exponent, and indicates there are two factors of 7. In the expression $7^3$, the 3 indicates that there are three factors of 7.

$$(7^3 = 7 \cdot 7 \cdot 7).$$

$3^5$ is a multiplication expression involving the exponent 5. $3^5$ is a multiplication expression in which 3 is used as a factor 5 times. $3^5 = 3 \cdot 3 \cdot 3 \cdot 3 \cdot 3$, and has an evaluation of 243.

To evaluate $(^-2)^4$, the following steps are used:

$$(^-2)^4 = ^-2 \cdot ^-2 \cdot ^-2 \cdot ^-2 = [(^-2 \cdot ^-2) \cdot ^-2] \cdot ^-2 = [4 \cdot ^-2] \cdot ^-2 = ^-8 \cdot ^-2 = 16$$

EXERCISE 3.10.1

1. What are the factors of $(^-3 \cdot 5) \cdot ^-10$?
2. Is 3 a factor of $5^3$?
3. What integer is a factor of $(^-3)^7$?
4. How many factors of 3 are there in the expression, $3^6$?
5. Evaluate:
   (a) $9^2$
   (b) $(^-3)^4$
   (c) $5^3$
   (d) $2^7$
   (e) $(^-1)^5$
   (f) $(^-4)^4$
   (g) $10^6$

### 3.11   Evaluating Numerical Expressions Involving both Addition and Multiplication

$5(7 + {}^-3)$ is a numerical expression[1] involving operations of both addition and multiplication. The parentheses indicate that the addition $7 + {}^-3$ is to be evaluated first, as follows:

$$5(7 + {}^-3)$$

$$5 \cdot 4$$

$$20$$

$9 + 3 \cdot {}^-5$ is also a numerical expression involving both addition and multiplication. When there are no parentheses to indicate which operation is to be performed first, an agreement always to be followed in mathematics is to perform the operation multiplication first, then any addition that is necessary to complete the evaluation. The evaluation of $9 + 3 \cdot {}^-5$ is therefore, done as follows:

$$9 + 3 \cdot {}^-5$$

$$9 + {}^-15$$

$$^-6$$

$^-4 \cdot {}^-2 - 5(6 + {}^-8)$ is evaluated as follows:

$$^-4 \cdot {}^-2 - 5(6 + {}^-8)$$

$$^-4 \cdot {}^-2 - 5 \cdot {}^-2$$

$$8 - {}^-10$$

$$8 + 10$$

$$18$$

EXERCISE 3.11.1

Evaluate:

1. $6(8 + {}^-12)$
2. $4 + 6 \cdot 3$
3. $(5 + 2) \cdot {}^-7$
4. $^-2 \cdot {}^-3 + 4 \cdot {}^-7$
5. $^-4(5 + 3 \cdot {}^-2)$

6. $^-8 \cdot 3 - 6$
7. $7 - 2 \cdot {}^-9$
8. $^-4 \cdot 6 - 5 \cdot 3$
9. $10 - 6(3 + {}^-8)$
10. $5(6 \cdot {}^-4 - 3)$

[1] $5(7 + {}^-3)$ is equivalent to $5 \cdot (7 + {}^-3)$. In both cases, the last operation to be performed is multiplication.

11. $5 \cdot {}^-1 - 2(6+3)$

12. $4(8-3) - 7$

13. ${}^-15 - 4 \cdot {}^-3$

14. $16 + 2 \cdot {}^-8$

15. $5(2 - 6 \cdot 3) - 3 \cdot 7$

16. ${}^-8(4 \cdot {}^-2 + 10)$

17. $(5 \cdot {}^-4 + 18) \cdot (6 - 3 \cdot 8)$

18. ${}^-3(4-9) - 6$

19. $2(8 + {}^-1) - 5({}^-6 + 2)$

20. $7(3 \cdot 4 - 9) - 2(2 \cdot 4 + 1)$

## 3.12 True and False Mathematical Statements

$2 \cdot 3 + 7 = 3 + 5 \cdot 2$ is a mathematical statement because it can be judged to be true or false by evaluating the numerical expressions on each side of the equal sign.

$$2 \cdot 3 + 7 = 3 + 5 \cdot 2$$

$$6 + 7 = 3 + 10$$

$$13 = 13$$

Since $13 = 13$ is a true statement, then $2 \cdot 3 + 7 = 3 + 5 \cdot 2$ is also a true statement.

$4(7+5) = 4 \cdot 7 + 5$ is another mathematical statement. Its truth or falsity can be determined by evaluating the numerical expressions on each side of the equal sign.

$$4(7+5) = 4 \cdot 7 + 5$$

$$4 \cdot 12 = 28 + 5$$

$$48 = 33$$

Since $48 = 33$ is a false statement, then $4(7+5) = 4 \cdot 7 + 5$ is also a false statement.

EXERCISE 3.12.1

Determine the truth or falsity of the following statements.

1. $4 + {}^-7 = {}^-7 + 4$

2. $5 \cdot {}^-8 = {}^-8 \cdot 5$

3. $6({}^-9 + 4) = 6 \cdot {}^-9 + 6 \cdot 4$

4. $8({}^-3 \cdot {}^-2) = (8 \cdot {}^-3) \cdot {}^-2$

5. $(16 + {}^-6) + {}^-3 = 16 + ({}^-6 + {}^-3)$

6. $4 \cdot {}^-7 + 4 \cdot {}^-8 = 4({}^-7 + {}^-8)$

7. $6(5-3) - 2(5-3) = 4(5-3)$

8. $5[18 - 3 \cdot 7] + 0 = 5[18 - 3 \cdot 7]$

9. $1[(13 - 7 \cdot 11) + {}^-43] = [(13 - 7 \cdot 11) + {}^-43]$

10. $17 \cdot 3 - 9 \cdot 3 = 8 \cdot 3$

11. ${}^-19 \cdot 6 + 4 \cdot 6 = {}^-15 \cdot 6$

12. $(2 \cdot {}^-7 + 6) + {}^-3 \cdot 4 = ({}^-3 \cdot 4 + 6) + 2 \cdot {}^-7$

### 3.13 Evaluating Open Expressions

$4 + 7 \cdot {}^-6$ is a numerical expression and names a unique integer, $^-38$. $2x - 5$ is an open expression in which the variable $x$ may be replaced by any integer. Whenever $x$ is replaced by an integer in $2x - 5$, a numerical expression is obtained.

To evaluate $2x - 5$ when $x = 7$, the following steps are used:

$$2x - 5, \quad \text{and} \quad x = 7$$
$$2 \cdot 7 - 5$$
$$14 - 5$$
$$14 + {}^-5$$
$$9$$

To evaluate $2x - 5$ when $x = 0$, the following steps are used:

$$2x - 5, \quad \text{and} \quad x = 0$$
$$2 \cdot 0 - 5$$
$$0 - 5$$
$$0 + {}^-5$$
$${}^-5$$

$3x + 7y$ is an open expression with two variables, $x$ and $y$. Each variable may be replaced by any integer. To evaluate $3x + 7y$ when $x = {}^-6$ and $y = 4$, the following steps are used:

$$3x + 7y, \quad \text{and} \quad x = {}^-6 \quad \text{and} \quad y = 4$$
$$3 \cdot {}^-6 + 7 \cdot 4$$
$${}^-18 + 28$$
$${}^-18 + (18 + 10)$$
$$10$$

To evaluate $3x + 7y$ when $x = 2$ and $y = {}^-3$, the following steps are used:

$$3x + 7y, \quad \text{and} \quad x = 2 \quad \text{and} \quad y = {}^-3$$
$$3 \cdot 2 + 7 \cdot {}^-3$$
$$6 + {}^-21$$
$$6 + ({}^-6 + {}^-15)$$
$${}^-15$$

EXERCISE 3.13.1

Evaluate:

1. $3 - 5x$, when $x = 2$.
2. $4x + 7$, when $x = {}^-5$.
3. $3x + 2(x - 3)$, when $x = 4$.
4. $x(x - 2)$, when $x = {}^-2$.
5. ${}^-7 + {}^-3y$, when $y = 4$.
6. $14c + 11$, when $c = {}^-1$.
7. $3(2r)$, when $r = {}^-4$.
8. ${}^-5(m + 2)$, when $m = {}^-6$.
9. $6a + 3b$, when $a = 5$ and $b = {}^-4$.
10. $4c - 6d$, when $c = 2$ and $d = 8$.
11. $7 + xy$, when $x = 3$ and $y = 5$.
12. $ab - 9$, when $a = {}^-4$ and $b = 3$.
13. ${}^-4 - xy$, when $x = {}^-5$ and $y = 2$.
14. $5x - 3y + 2z$, when $x = 4, y = 2$, and $z = {}^-1$.
15. $3x - yz$, when $x = 8, y = 2$, and $z = 3$.
16. $x^2$, when $x = 6$.
17. $y^3$, when $y = {}^-3$.
18. $z^2$, when $z = {}^-7$.
19. $x^2 \cdot x^3$, when $x = 2$.
20. $y \cdot y^2$, when $y = {}^-4$.

### 3.14  Open Sentences

$4x + yx = x(4 + y)$ is an open sentence with two variables, $x$ and $y$. Each of the variables may be replaced by any integer. When the variables are replaced by integers a mathematical statement is obtained. The mathematical statement may be either true or false.

For the open sentence, $4x + yx = x(4 + y)$, if $x = 7$ and $y = {}^-3$, the following mathematical statement is obtained.

$$4x + yx = x(4 + y)$$

$$4 \cdot 7 + {}^-3 \cdot 7 = 7(4 + {}^-3)$$

By evaluating the numerical expressions on both sides of the equal sign, the truth or falsity of the mathematical statement can be determined, as follows:

$$4 \cdot 7 + {}^-3 \cdot 7 = 7(4 + {}^-3)$$

$$28 + {}^-21 = 7 \cdot 1$$

$$7 = 7$$

When $x = 7$ and $y = {}^-3$, the open sentence $4x + yx = x(4 + y)$ becomes a true statement.

For the open sentence, $5x - 3x = {}^-8x$, if $x = 4$, the following mathematical statement is obtained:

$$5x - 3x = {}^-8x$$
$$5 \cdot 4 - 3 \cdot 4 = {}^-8 \cdot 4$$

By evaluating the numerical expressions on both sides of the equal sign, the truth or falsity of the mathematical statement can be determined, as follows:

$$5 \cdot 4 - 3 \cdot 4 = {}^-8 \cdot 4$$
$$20 - 12 = {}^-32$$
$$20 + {}^-12 = {}^-32$$
$$8 = {}^-32$$

$8 = {}^-32$ is false. Therefore, when $x = 4$, the open sentence $5x - 3x = {}^-8x$, becomes a false statement.

EXERCISE 3.14.1

Replace each variable by the given value and determine whether the resulting mathematical statement is true or false.

1. $4 + x = x + 4$, when $x = {}^-9$.
2. $(6 + 3x) - 7 = 6 + (3x - 7)$, when $x = {}^-2$.
3. ${}^-4x + 9 = 5x$, when $x = 3$.
4. $4 - (x + 2) = (4 - x) + 2$, when $x = 5$.
5. ${}^-3(4x) = {}^-12x$, when $x = {}^-2$.
6. ${}^-7(x - 2) = {}^-7x + 14$, when $x = 5$.
7. $(9 + 3x) + {}^-2 = 3x + 7$, when $x = {}^-5$.
8. ${}^-(x + {}^-3) = {}^-x + 3$, when $x = 4$.
9. $2(x - 5) = 2x - 5$, when $x = 4$.
10. ${}^-6x - x = {}^-7x$, when $x = 3$.
11. $x + y = y + x$, when $x = 4$ and $y = {}^-11$.
12. $xy = yx$, when $x = {}^-7$ and $y = 6$.
13. $3x + 2y = 5xy$, when $x = {}^-2$ and $y = 4$.
14. $(x + 3y) + z = x + (3y + z)$, when $x = 3, y = {}^-2$, and $z = 6$.
15. $x(y - 3) = xy - 3x$, when $x = {}^-5$ and $y = 4$.
16. $5xy = y \cdot (5x)$, when $x = {}^-1$ and $y = {}^-3$.
17. $3x - (x - 2y) = 2x + 2y$, when $x = 3$ and $y = 4$.
18. ${}^-(2x - 3y) = {}^-2x + 3y$, when $x = 7$ and $y = 2$.
19. $(2x + 7y) + (3x - 4y) = 5x + 3y$, when $x = {}^-3$ and $y = {}^-2$.
20. $x^3 \cdot x^2 = x^5$, when $x = 2$.

### 3.15    Laws of Addition for the Integers

The open sentence $x + y = y + x$ becomes a true statement for any integer replacements of $x$ and $y$. This fact is stated as the Commutative Law for Addition of Integers.

*The Commutative Law for Addition*

> The sum of any two integers is the same regardless of the order in which they are added.

The Commutative Law of Addition is the basis for claiming that each of the following open sentences will result in a true mathematical statement for any integer replacements of the variables.

$$x + 4 = 4 + x$$

$$2x + 5y = 5y + 2x$$

$$-3x + 2 = 2 - 3x$$

$$(3x + 5y) + 6z = 6z + (3x + 5y)$$

$$8x - 3y = -3y + 8x$$

A second law of addition is based upon the open sentence, $(x + y) + z = x + (y + z)$. This open sentence will become a true mathematical statement for any integer replacements of the variables.

*The Associative Law for Addition*

> The addition of three integers is the same whether the sum of the first two is added to the third or the first is added to the sum of the last two.

The Associative Law for Addition is the basis for claiming that each of the following open sentences will result in a true mathematical statement for any integer replacements of the variables. Notice that in each case the order of the terms remains the same, but the grouping has been altered.

$$(x + 9) + -3 = x + (9 + -3)$$

$$(4 - 3x) + 5x = 4 + (-3x + 5x)$$

$$(2x - 5) + (4 - 7x) = [(2x - 5) + 4] - 7x$$

$$(3x - 7y) + 9y = 3x + (-7y + 9y)$$

Although the Commutative and Associative Laws of Addition of integers had their counterparts in the set of counting numbers, the third law of addition of integers is dependent upon zero, which is not a counting number.

*Zero is the Identity Element for Addition*

The sum of any integer and zero is that same integer.

Each of the following open sentences will become a true mathematical statement for any integer replacements of the variables because zero is the identity element for addition.

$$x + 0 = x$$

$$^-17y + 0 = {}^-17y$$

$$(4x - 3y) + 0 = 4x - 3y$$

$$2x + (9 + {}^-9) = 2x$$

*Every Integer Has an Opposite*

Given any integer there is another integer such that their sum is zero, $x + {}^-x = 0$ for all integers $x$.

EXERCISE 3.15.1

For each open sentence, state the law of addition which claims the sentence will become a true mathematical statement for any integer replacements of the variables.

1. $6a + c = c + 6a$
2. $14z = 14z + 0$
3. $(5p + 3r) + 2s = 5p + (3r + 2s)$
4. $(6z - 9) + 3 = 6z + (^-9 + 3)$
5. $(4r + 3) - 3r = {}^-3r + (4r + 3)$
6. $[\,13z + {}^-18x) - 5y\,] + 0 = (13z + {}^-18x) - 5y$
7. $2x - 4 = {}^-4 + 2x$
8. $(5 + 3x) - 11x = 5 + (3x - 11x)$
9. $7x + (^-4 + 4) = 7x$
10. $(5x - 3) + (^-2x - 6) = [\,(5x - 3) - 2x\,] - 6$

### 3.16  Simplifying Addition Expressions

The four laws for addition for the integers are used to simplify many expressions.

The open expressions $(4 + 2x) - 7$ and $2x - 3$ are equivalent, because $(4 + 2x) - 7 = 2x - 3$ will become a true mathematical statement for any integer replacements for $x$. The following steps are used to simplify $(4 + 2x) - 7$:

$$(4 + 2x) - 7$$
$$(2x + 4) - 7$$
$$2x + (4 - 7)$$
$$2x + (4 + {}^-7)$$
$$2x + {}^-3$$
$$2x - 3$$

The simplification of $(3 - 5x) - 3$ is done as follows:

$$(3 - 5x) - 3$$
$$(3 + {}^-5x) + {}^-3$$
$$({}^-5x + 3) + {}^-3$$
$${}^-5x + (3 + {}^-3)$$
$${}^-5x + 0$$
$${}^-5x$$

EXERCISE 3.16.1

Simplify each of the following open expressions:

1.  $(8 + 3x) + 9$
2.  $(3 - 4x) + 2$
3.  $(3x - 7) + 5$
4.  $(2 + x) - 6$
5.  $4 + (3x - 7)$
6.  $(2x - 6) - 1$
7.  $^-5 + (6 - 6x)$
8.  $(4x - 8) + 8$
9.  $(9 - 3x) - 2$
10. $(x - 10) - 6$

11. $(5 + 8x) - 7$
12. $(3x - 8) - 1$
13. $(4 - 7x) + 6$
14. $5 + (3x - 2)$
15. $^-6 + (5x - 12)$
16. $^-7 + (5 - 6x)$
17. $(2x - 3) + 8$
18. $[2 + (6x - 5)] - 9$
19. $3 + [(2 - 3x) - 7]$
20. $[17 + (3 - 3x)] - 8$

### 3.17   Laws of Multiplication for the Integers

There are four laws of multiplication to be studied in this section. The first of these laws has a counterpart in the set of counting numbers.

$xy = yx$ becomes a true mathematical statement for any integer replacements of the variables $x$ and $y$. This fact is known as the Commutative Law for Multiplication.

*The Commutative Law for Multiplication of Integers*

The product of two integers is the same regardless of the order in which they are multiplied.

The Commutative Law for Multiplication is the basis for claiming that each of the following open sentences will become a true mathematical statement for any integer replacements of the variables.

$$^-5y = y \cdot {}^-5$$

$$^-8(3x) = (3x) \cdot {}^-8$$

$$(4 + x) \cdot {}^-2 = {}^-2(4 + x)$$

$(xy)z = x(yz)$ becomes a true mathematical statement for any integer replacements of the variables $x$, $y$, and $z$. This fact is known as the Associative Law for Multiplication.

*The Associative Law for Multiplication of Integers*

The product of three integers is the same whether the product of the first two is multiplied by the third or the first integer is multiplied by the product of the last two.

The Associative Law for Multiplication is the basis for claiming that each of the following open sentences will become true mathematical statements for any replacements of the variables.

$$(3x)y = 3(xy)$$

$$^-4(2x) = ({}^-4 \cdot 2)x$$

$$(5 \cdot {}^-3)z = 5({}^-3z)$$

Positive one is the identity element for multiplication. $1x = x$ will become a true statement for any integer replacement for $x$.

*One is the Identity Element for Multiplication*

The product of one and any integer is that same integer.

Each of the following sentences will become a true mathematical statement for any replacement of the variables because one is the identity element for multiplication:

$$^-5x = 1(^-5x)$$

$$1z = z$$

$$(^-4 + 3x) = 1(^-4 + 3x)$$

$$^-9x + x = ^-9x + 1x$$

The last multiplication property for the integers is concerned with the integer, $^-1$, and its relationship between any integer and its opposite. The opposite of 5 is $^-5$, which can be shown in another way by using the equality $^-1 \cdot 5 = ^-5$. The opposite of $^-8$ is 8, but another way of showing the relationship is the equality $^-1 \cdot ^-8 = 8$. If $^-x$ is read, "the opposite of $x$," the open sentence $^-1x = ^-x$ will become a true mathematical statement for any integer replacement of the variable $x$.

*The Multiplication Law of $^-1$*

The product of negative one and any integer is the opposite of that integer.

Each of the following open sentences will become a true mathematical statement for any integer replacements of the variables, because of the multiplication property of $^-1$:

$$^-z = ^-1z$$

$$4x + ^-x = 4x + ^-1x$$

$$^-(3x - 2) = ^-1(3x - 2)$$

$$(5x) \cdot (^-y) = (5x) \cdot (^-1y)$$

EXERCISE 3.17.1

For each open sentence, state the law of multiplication which claims that it will become a true mathematical statement for all replacements of the variables.

1.  $^-4c = c \cdot ^-4$
2.  $^-5(8w) = (^-5 \cdot 8)w$
3.  $r = 1r$

4.  $^-1q = ^-q$
5.  $(4x - 3) = 1(4x - 3)$
6.  $(^-2z)y = y(^-2z)$

7.  $^-(3x - 8) = ^-1(3x - 8)$        9.   $^-4x + x = ^-4x + 1x$
8.  $(^-4x)y = ^-4(xy)$              10.  $^-x + 17 = ^-1x + 17$

### 3.18  Simplifying Multiplication Expressions

The four laws for multiplication for integers are used to simplify many open expressions.

The open expressions, $(^-2x) \cdot 7$ and $^-14x$, are equivalent expressions because $(^-2x) \cdot 7 = ^-14x$ will become a true mathematical statement for any integer replacement of the variable $x$. The steps in simplifying $(^-2x) \cdot 7$, are as follows:

$$(^-2x) \cdot 7$$

$$7(^-2x)$$

$$(7 \cdot {}^-2)x$$

$$^-14x$$

To simplify $(^-5x) \cdot (2y)$, the following steps are used:

$$(^-5x) \cdot (2y)$$

$$(^-5 \cdot 2) \cdot (xy)$$

$$^-10xy$$

To simplify $x \cdot (3y) \cdot {}^-z$, the following steps are used:

$$x \cdot (3y) \cdot {}^-z$$

$$1x \cdot (3y) \cdot {}^-1z$$

$$(1 \cdot 3 \cdot {}^-1) \cdot (xyz)$$

$$^-3xyz$$

EXERCISE 3.18.1

Simplify each of the following open expressions:

1.  $^-3(4x)$                    11.  $^-6(5x)$
2.  $^-2(^-9z)$                  12.  $^-4(^-3x)$
3.  $(^-5k) \cdot 3$             13.  $(^-7x) \cdot {}^-1$
4.  $(4m) \cdot {}^-1$           14.  $^-5(3x)$
5.  $(^-6x) \cdot (7y)$          15.  $(4x) \cdot {}^-8$
6.  $(^-3a) \cdot (^-c)$         16.  $(^-5x) \cdot {}^-9$
7.  $^-x \cdot (2y) \cdot z$     17.  $(^-2x) \cdot (7y)$
8.  $(^-2x) \cdot (5y) \cdot (^-3z)$   18.  $(^-3x) \cdot (^-8y)$
9.  $(^-5x) \cdot (^-8y) \cdot z$      19.  $(^-2x) \cdot (^-5y) \cdot z$
10.  $4x \cdot (^-3y) \cdot (^-2z)$    20.  $x \cdot (^-2y) \cdot z$

### 3.19   The Distributive Law of Multiplication over Addition

The Distributive Law of Multiplication over Addition is the only law concerned with both multiplication and addition of integers.

$$5(2 + 3) = 5 \cdot 2 + 5 \cdot 3$$

$$^-6(4 + {}^-7) = {}^-6 \cdot 4 + {}^-6 \cdot {}^-7$$

$$2({}^-8 + {}^-3) = 2 \cdot {}^-8 + 2 \cdot {}^-3$$

The preceding examples are three true statements in which the first operation performed on the left side of the equality is addition, but the first operation performed on the right side of the equality is multiplication. In fact, the distributive property is concerned with a special type of expression in which the order of the operations (addition or multiplication) may be changed.

*The Distributive Law of Multiplication over Addition*

The multiplication of an integer by the sum of two integers is the same as the sum of the products of the first integer with each of the other two integers.

The open sentence $x(y + z) = xy + xz$ becomes a true mathematical statement for any integer replacements of $x$, $y$, and $z$, because of the Distributive Law of Multiplication over Addition.

The parentheses of the open expression $5(x - 3)$ may be removed by using the distributive property as follows:

$$5(x - 3)$$

$$5(x + {}^-3)$$

$$5x + 5 \cdot {}^-3$$

$$5x + {}^-15$$

$$5x - 15$$

Using the distributive property, the parentheses of $^-2(3x - 7)$ may be removed as follows:

$$^-2(3x - 7)$$

$$^-2(3x + {}^-7)$$

$$^-2 \cdot 3x + {}^-2 \cdot {}^-7$$

$$^-6x + 14$$

$^-(5x - 4)$ is simplified as follows:

$$^-(5x - 4)$$
$$^-1(5x + ^-4)$$
$$^-1 \cdot 5x + ^-1 \cdot ^-4$$
$$^-5x + 4$$

Expressions such as $^-4x + 7x$, $^-3x - x$, and $x - 9x$ are simplified by using the distributive property in the following manner:

(a) $^-4x + 7x$

$(^-4 + 7)x$

$3x$

(b) $^-3x - x$

$^-3x + ^-x$

$^-3x + ^-1x$

$(^-3 + ^-1)x$

$^-4x$

(c) $x - 9x$

$1x + ^-9x$

$(1 + ^-9)x$

$^-8x$

EXERCISE 3.19.1

1. Simplify each expression by removing the parentheses.
   (a) $3(x + 2)$
   (b) $^-5(x - 4)$
   (c) $^-2(3x + 1)$
   (d) $5(4 - 3x)$
   (e) $^-(2x - 9)$
   (f) $^-(5 + 7x)$
   (g) $4(3x - 2)$
   (h) $^-5(9 + 6x)$
   (i) $4(7 - x)$
   (j) $^-9(4x - 1)$
2. Simplify each expression by using the distributive property.
   (a) $5x + ^-3x$
   (b) $7x - 9x$
   (c) $^-8y + 4y$
   (d) $^-6w + w$
   (e) $5k - k$
   (f) $^-3x - 5x$
   (g) $7z + 8z$
   (h) $^-2x - x$
   (i) $4r - 11r$
   (j) $w - 2w$

3. Simplify each expression by removing the parentheses.

(a) $5(x - 7)$

(b) $^-2(3x + 8)$

(c) $^-(4x - 6)$

(d) $^-5(6 + 2x)$

(e) $7(4 - 8x)$

(f) $^-(5x + 1)$

(g) $^-(8x - 3)$

(h) $3(5x - 9)$

(i) $^-7(2 - 3x)$

(j) $6(9x - 7y)$

## 3.20  Simplifying Open Expressions Involving both Addition and Multiplication

The laws of the integers make it possible to simplify rather involved open expressions such as $2(x - 5) - 3(2x - 6)$. The first step in simplifying this type of open expression is to remove the parentheses by using the distributive law. The second step in the simplification is to properly group the addends of the expression. The simplification of $2(x - 5) - 3(2x - 6)$ is done as follows:

$$2(x - 5) - 3(2x - 6)$$

$$2(x + {}^-5) + {}^-3(2x + {}^-6)$$

$$2x + 2 \cdot {}^-5 + {}^-3 \cdot 2x + {}^-3 \cdot {}^-6$$

$$2x + {}^-10 + {}^-6x + 18$$

$$(2x + {}^-6x) + ({}^-10 + 18)$$

$$(2 + {}^-6)x + 8$$

$${}^-4x + 8$$

$(2x - 7) - (5 - 4x)$ is simplified as follows:

$$(2x - 7) - (5 - 4x)$$

$$(2x + {}^-7) + {}^-(5 + {}^-4x)$$

$$1(2x + {}^-7) + {}^-1(5 + {}^-4x)$$

$$2x + {}^-7 + {}^-5 + 4x$$

$$(2x + 4x) + ({}^-7 + {}^-5)$$

$$(2 + 4)x + {}^-12$$

$$6x - 12$$

EXERCISE 3.20.1

Simplify the following open expressions:

1.  $4(x + 2) - 3(x - 1)$
2.  $2(3x - 1) + 4(3 - x)$
3.  $5(x - 3) - 2(x - 1)$
4.  $2(5x - 2) + 3(x + 2)$
5.  $3(2x - 7) - (x - 3)$
6.  $5(x - 8) - (2x + 4)$
7.  $4(3x + 1) - 2(6x - 5)$
8.  $2(4x - 1) + (2x + 1)$
9.  $(3x - 7) - (8x + 3)$
10. $6(x - 5) - 3(5x + 2)$

11. $3(x - 7) - 2(x + 3)$
12. $4(2x + 1) - (5x - 2)$
13. $3(6x - 7) + (x - 9)$
14. $(2x - 9) - 5(x - 1)$
15. $5(7x - 3) - (13x - 9)$
16. $^{-}3(x + 8) - 2(x - 9)$
17. $2(6x - 1) - 3(4x + 1)$
18. $(5x + 7) - (2x - 6)$
19. $6(x - 1) - (3x + 5)$
20. $2(3x + 4) - (5x + 6)$

## 3.21  Simplifying Multiplication Expressions Using Exponents

$3^5$ is a numerical expression using the exponent 5. $3^5$ means $3 \cdot 3 \cdot 3 \cdot 3 \cdot 3$, where the exponent 5 indicates that 3 is to be used as a factor five times in evaluating the multiplication expression.

$x^4$ is an open expression. $x^4$ means $x \cdot x \cdot x \cdot x$ or $xxxx$. The exponent 4 indicates that $x$ is to be used as a factor four times in the expression.

$x^2 \cdot x^3$ is an open expression involving $x^2$ and $x^3$. To simplify $x^2 \cdot x^3$, the following steps are used:

$$x^2 \cdot x^3$$

$$xx \cdot xxx$$

$$xxxxx$$

$$x^5$$

$x \cdot x^6$ is simplified as follows:

$$x \cdot x^6$$

$$x \cdot xxxxxx$$

$$xxxxxxx$$

$$x^7$$

EXERCISE 3.21.1

Simplify the following open expressions:

1.  $x^4 \cdot x^2$
2.  $x^3 \cdot x^5$

3.  $x^2 \cdot x^6$
4.  $x \cdot x^3$

5. $x^2 \cdot x$

6. $x \cdot x$

7. $x^5 \cdot x^7$

8. $x^{12} \cdot x^3$

9. $x^{21} \cdot x^9$

10. $x^{43} \cdot x^6$

## 3.22  Truth Sets for Simple Equations

$x + 9 = 4$ is an open sentence or equation. If $\{-8, -5, -3, 1\}$ is the domain of the variable $x$, then to solve $x + 9 = 4$, it is necessary to find any elements of the domain that make $x + 9 = 4$ a true statement.

If   $x = -8$,   $-8 + 9 = 4$   is false.

If   $x = -5$,   $-5 + 9 = 4$   is true.

If   $x = -3$,   $-3 + 9 = 4$   is false.

If   $x = 1$,    $1 + 9 = 4$   is false.

Hence, $\{-5\}$ is the truth set for $x + 9 = 4$.

$-3x = 12$ is an open sentence or equation. If $\{-8, -6, -4, -2\}$ is used as the domain for the variable $x$, the following mathematical statements are obtained:

If   $x = -8$,   $-3 \cdot -8 = 12$   is false.

If   $x = -6$,   $-3 \cdot -6 = 12$   is false.

If   $x = -4$,   $-3 \cdot -4 = 12$   is true.

If   $x = -2$,   $-3 \cdot -2 = 12$   is false.

Hence, $\{-4\}$ is the truth set of $-3x = 12$.

Using $\{-7, -4, 1, 5\}$ as the domain for $x + 4 = 1$, the truth set is $\{\ \}$, because each replacement for $x$ results in a false statement.

EXERCISE 3.22.1

Each problem contains an equation and a domain set. Find the truth set.

1. $x + 7 = 5$,   $\{-9, -6, -2, 0, 4\}$

2. $x - 6 = -3$,   $\{-9, -6, 0, 3, 9\}$

3. $x + 8 = 11$,   $\{-9, -3, 0, 3, 9\}$

4. $x - 9 = -10$,   $\{-3, -1, 1, 3, 8\}$

5. $4x = -20$,   $\{-10, -5, 0, 5, 10\}$

6. $-6x = -18$,   $\{-12, -3, 3, 13\}$

7. $-7x = 42$,   $\{-12, -6, 6, 12\}$

8. $-3x = 7$,   $\{-3, -2, -1, 0, 1, 2, 3\}$

9. $2x = 10$,   $\{-8, -5, 0, 5, 8\}$

10. $-7x = 0$,   $\{-7, -1, 0, 1, 7\}$

11. $x + 8 = 3$, $\{-7, -5, 5, 8, 11\}$
12. $x - 9 = -2$, $\{11, -11, 7, -7, 0\}$
13. $x + 7 = -4$, $\{11, -11, 7, -3, 1\}$
14. $x - 6 = 5$, $\{11, -11, 7, 1, -1\}$
15. $x + 1 = -6$, $\{11, 5, -5, -7, 7\}$
16. $x - 3 = -8$, $\{11, 5, -5, -7, -11\}$
17. $4x = -20$, $\{11, 5, -5, -16, -24\}$
18. $-x = -5$, $\{11, 5, -5, 0, -1\}$
19. $-5x = 7$, $\{0, -1, -7, -2, 3\}$
20. $-3x = 12$, $\{4, -9, -4, -9, 0\}$

### 3.23 Using the Set of Integers as the Domain

For the remainder of this chapter the set of integers, $\{\ldots, -3, -2, -1, 0, 1, 2, 3, \ldots\}$ will be used as the domain for all equations. Using the set of integers as the domain, the truth set of $x + 8 = 2$ is $\{-6\}$, because $-6$ is the only integer that can be added to 8 to produce a sum of 2.

The truth set of $8x = -56$ is $\{-7\}$, because $-7$ is the only integer that can be multiplied by 8 to produce a product of $-56$.

The truth set of $-5x = 21$ is $\{\quad\}$, because there is no integer that can be multiplied by $-5$ to produce a product of 21.

The truth set of $(4x - 5) - 3x = 8$ is not so obvious. To find the truth set of $(4x - 5) - 3x = 8$, the first step is to simplify the expression, $(4x - 5) - 3x$, as shown in the following steps:

$$(4x - 5) - 3x = 8$$

$$(4x + {}^-5) + {}^-3x = 8$$

$$(4x + {}^-3x) + {}^-5 = 8$$

$$x + {}^-5 = 8$$

Since $(4x - 5) - 3x$ and $x + {}^-5$ are equivalent, any integer replacement for $x$ that makes $x + {}^-5 = 8$ a true statement will also make $(4x - 5) - 3x = 8$ a true statement. Hence, $\{13\}$ is the truth set of $(4x - 5) - 3x = 8$, because 13 is the only integer that makes $x + {}^-5 = 8$ a true statement.

To find the truth set of $9 - (5x + 9) = 35$, it is necessary to simplify the left member of the equation, as follows:

$$9 - (5x + 9) = 35$$

$$9 - 1(5x + 9) = 35$$

$$9 - 5x - 9 = 35$$

$$-5x = 35$$

Therefore, $\{^-7\}$ is the truth set of $9 - (5x + 9) = 35$, because $^-7$ is the only integer that makes $^-5x = 35$ a true statement.

The truth set of $7 - (3x + 7) = 10$ is the empty set, $\{\quad\}$.

$$7 - (3x + 7) = 10$$

$$7 - 3x - 7 = 10$$

$$^-3x = 10$$

There is no integer which can be multiplied by $^-3$ to give 10.

EXERCISE 3.23.1

Find the truth set of each equation by first simplifying the left member of the equation.

1. $(5x - 7) - 4x = 2$
2. $4x + (5 - 3x) = 2$
3. $4 + (9x - 4) = 18$
4. $8 - (3x + 8) = 12$
5. $2(3x - 7) - 6x = 1$
6. $12 - 3(x + 4) = {}^-21$
7. $11x - 5(2x - 3) = 15$
8. $3(x + 1) - 2(x - 6) = 1$
9. $6(x - 2) + 4(x + 3) = {}^-20$
10. $2(3x - 7) - 5(x - 2) = 4$
11. $(7x - 3) - 6x = 4$
12. $5x - (4x - 1) = {}^-6$
13. $(5x - 9) + 9 = {}^-10$
14. $(2 - 4x) - 2 = 8$
15. $^-5x + (6x - 3) = 4$
16. $(5 - 8x) - 5 = {}^-20$
17. $3(2x - 4) - 2(x - 6) = {}^-12$
18. $5(x - 2) - (4x - 3) = 6$
19. $(x - 8) - 2(x - 4) = {}^-7$
20. $5(2x - 1) - 3(3x - 7) = 6$

## 3.24  Solving Equations by Adding Opposites

The equations, $x + 4 = 3$ and $(x + 4) + 7 = 3 + 7$, have the same truth set, $\{^-1\}$. These two equations have the same truth set because 7 was added to both members of the equation, $x + 4 = 3$, to obtain the second equation, $(x + 4) + 7 = 3 + 7$. Any integer may be added to both members of an equation to obtain a second equation that will have the same truth set as the first. Following are three pairs of equations; in each case the second equation was obtained by adding the same integer to both sides of the first equation.

$$2x - 3 = 7 \quad \text{and} \quad (2x - 3) + 8 = 7 + 8$$

$$4x + 5 = 21 \quad \text{and} \quad (4x + 5) + {}^-3 = 21 + {}^-3$$

$$5x + 7 = 27 \quad \text{and} \quad (5x + 7) + {}^-7 = 27 + {}^-7$$

The last pair of equations is of special interest. $5x + 7 = 27$ and $(5x + 7) + {}^-7 = 27 + {}^-7$ have the same truth set. By simplifying both sides of the second equation, $(5x + 7) + {}^-7 = 27 + {}^-7$, we obtain $5x = 20$, which has a truth set of $\{4\}$. Since 4 is the only integer that makes the second equation a true statement, and the first and second equations are equivalent, $\{4\}$ is also the truth set of the first equation, $5x + 7 = 27$.

Every integer has an opposite. This property makes it possible to find the truth set of $2x - 9 = 7$, by first adding the opposite of $^-9$ to both sides of the equation, as follows:

$$2x - 9 = 7$$

$$(2x - 9) + 9 = 7 + 9$$

$$2x + ({}^-9 + 9) = 16$$

$$2x + 0 = 16$$

$$2x = 16$$

The truth set of $2x - 9 = 7$ is $\{8\}$, because 8 is the only integer that makes $2x = 16$ a true statement.

EXERCISE 3.24.1

1. Find the truth set of $5x + 8 = 3$ by first adding $^-8$ to both sides of the equation.
2. Find the truth set of $3x - 5 = 10$ by first adding 5 to both sides of the equation.
3. Find the truth set of $^-6x + 2 = 26$ by first adding the opposite of 2 to both sides of the equation.
4. Find the truth set of $4x - 7 = 25$ by first adding the opposite of $^-7$ to both sides of the equation.

EXERCISE 3.24.2

Find the truth set of each equation by adding the appropriate number to both sides of the equation.

| | | | |
|---|---|---|---|
| 1. | $2x + 5 = 11$ | 11. | $3x - 14 = 1$ |
| 2. | $3x - 7 = 23$ | 12. | $7x + 23 = 2$ |
| 3. | $2x - 9 = 1$ | 13. | $5x - 9 = 31$ |
| 4. | $4x + 9 = 5$ | 14. | $4x + 7 = 7$ |
| 5. | $^-3x + 2 = 14$ | 15. | $3 + 2x = 15$ |
| 6. | $^-8x - 5 = 3$ | 16. | $^-8 + 3x = 10$ |
| 7. | $5x + 12 = ^-13$ | 17. | $2x + 3 = 12$ |
| 8. | $4x + 10 = 41$ | 18. | $4 - 2x = 10$ |
| 9. | $10x - 3 = 47$ | 19. | $7 - 3x = 40$ |
| 10. | $^-6x + 5 = 29$ | 20. | $32 - 5x = 17$ |

The variable $x$ appears on both sides of the equation, $5x - 4 = 2x + 17$. The opposite of $2x$ is $^-2x$, and the opposite of $^-4$ is 4. These opposites are used in solving the equation, $5x - 4 = 2x + 17$, as shown in the following steps:

$$5x - 4 = 2x + 17$$

$$(5x - 4) + {}^-2x = (2x + 17) + {}^-2x$$

$$3x - 4 = 17$$

$$(3x - 4) + 4 = 17 + 4$$

$$3x + 0 = 21$$

$$3x = 21$$

The truth set of $5x - 4 = 2x + 17$ is $\{7\}$ because 7 is the only integer that will make $3x = 21$ a true statement.

The opposite of $^-9x$ is $9x$. This fact is used in finding the truth set of $4x - 2 = 24 - 9x$, as follows:

$$4x - 2 = 24 - 9x$$

$$(4x - 2) + 9x = (24 - 9x) + 9x$$

$$13x - 2 = 24$$

$$(13x - 2) + 2 = 24 + 2$$

$$13x + 0 = 26$$

$$13x = 26$$

The truth set of $4x - 2 = 24 - 9x$ is $\{2\}$ because 2 is the only integer that will make $13x = 26$ become a true statement.

EXERCISE 3.24.3

Find the truth sets by adding opposites.

1. $7x + 3 = 23 + 2x$
2. $2x - 7 = 5x + 2$
3. $6x + 9 = 1 - 2x$
4. $x + 7 = 22 - 4x$
5. $3x - 2 = 2x + 11$
6. $5x + 4 = 3x + 5$
7. $7x - 6 = 4x + 9$
8. $2x + 13 = 3 - 3x$
9. $6x - 9 = 13 - 5x$
10. $6 - 2x = 9 + x$

11. $5x - 2 = 3x - 14$
12. $6x + 3 = x + 13$
13. $2x - 7 = 4x + 5$
14. $7x + 3 = 5x - 4$
15. $3x + 2 = 10 - x$
16. $3 - 5x = 2x + 17$
17. $4x - 7 = 8x - 7$
18. $5 - 4x = 23 - x$
19. $x + 7 = 3x - 19$
20. $4 - 2x = 3x - 11$

Before adding opposites to both sides of any equation, it is important that the expressions on both sides of the equation be completely simplified. Both sides of the equation, $4 - 3(x + 2) = 6x - 2(4x + 5)$, should be simplified before attempting to add opposites to find the truth set.

$$4 - 3(x + 2) = 6x - 2(4x + 5)$$

$$4 - 3x - 6 = 6x - 8x - 10$$

$$^{-}3x - 2 = {}^{-}2x - 10$$

$$(^{-}3x - 2) + 2x = (^{-}2x - 10) + 2x$$

$$^{-}x - 2 = {}^{-}10$$

$$(^{-}x - 2) + 2 = {}^{-}10 + 2$$

$$^{-}x + 0 = {}^{-}8$$

$$^{-}x = {}^{-}8$$

Since $^{-}x$ is equivalent to $^{-}1x$, the truth set of $4 - 3(x + 2) = 6x - 2(4x + 5)$ is $\{8\}$, because 8 is the only integer that can make $^{-}1x = {}^{-}8$ a true statement.

EXERCISE 3.24.4

Find the truth set by first simplifying both sides of the equation.

1. $4x - 3(x - 2) = 2x - 9$
2. $7 - 2(3x + 1) = 4x - 15$

3.  $6 - 5(4 - x) = 2x - 4(x - 7)$
4.  $5x - (3x + 2) = 4 - (x - 8)$
5.  $14 + 3(x + 1) = 5x - 3(x + 9)$
6.  $5(3x - 7) + 2(x - 3) = 11x + 1$
7.  $8x - (3x + 1) = 2x + 4$
8.  $6(x + 4) - 2(5x + 1) = {}^-10$
9.  $5(3x - 7) - 8x = 5x + 1$
10. $7 - (6 - 2x) = 5x - 2(x + 6)$
11. $3x - (x + 5) = x + 12$
12. $9 - 3(x - 2) = 2x + 5$
13. $4x + 2(x - 6) = 3x - 9$
14. $6 + 3(4 - 2x) = 9 - 5x$
15. $5x - 2(x + 1) = 4 - (3 - 2x)$
16. $3x + 2(4 - 2x) = 5 - 2x$
17. $7x + (x - 3) = 2 - 3(x + 1)$
18. $2x - 5(x - 1) = x - 3(x + 3)$
19. $6 - (4 - x) = 5x - (x + 7)$
20. $3x - 2(x + 3) = 6 - (1 + 2x)$

### 3.25   Simple Inequalities

Figure 3.3 below shows graphically the set of integers, with the positive integers to the right of zero and the negative integers to the left of zero.

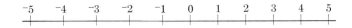

FIGURE 3.3

$x + 5 > 8$ is an inequality which will become a true statement for any integer to the right of 3 in figure 3.3. For example, if $x = 4$, then the statement $4 + 5 > 8$ is true because $4 + 5$ is greater than 8.

$x + 7 < 13$ is an inequality which will become a true statement for any integer to the left of 6 in Figure 3.3. For example, if $x = 3$, then the statement $3 + 7 < 13$ is true, because $3 + 7$ is less than 13.

Whether one integer is greater or less than another integer can be determined by using figure 3.3. In the figure, 3 is to the right of $^-5$ which tells us that 3 is greater (larger) than $^-5$. Similarly, 0 is greater than $^-4$ because 0 is to the right of $^-4$ in figure 3.3. $^-1$ is greater than $^-6$ because $^-1$ is to the right of $^-6$.

The same integer may be added to both sides of an inequality without altering the set of numbers that will make the inequality a true statement. For example, to simplify the inequality, $9x - 7 > 8x + 1$, the following steps are used:

$$9x - 7 > 8x + 1$$

$$(9x - 7) + {}^-8x > (8x + 1) + {}^-8x$$

$$x - 7 > 1$$

$$(x - 7) + 7 > 1 + 7$$

$$x + 0 > 8$$

$$x > 8$$

To simplify $7x + 5 < 6x - 4$ the following steps are used:

$$7x + 5 < 6x - 4$$

$$(7x + 5) + {}^-6x < (6x - 4) + {}^-6x$$

$$x + 5 < {}^-4$$

$$(x + 5) + {}^-5 < {}^-4 + {}^-5$$

$$x < {}^-9$$

EXERCISE 3.25.1

1. Simplify each of the inequalities by adding opposites to both sides of the inequality.
   (a) $4x - 3 > 2 + 3x$
   (b) $6 - 8x < 1 - 9x$
   (c) $5(x - 3) - 2x > 2x - 9$
   (d) $4x - (x + 1) < 2x + 7$
   (e) $7x - 5 > 6x + 1$
   (f) $4x - 3(x - 5) > 5$
   (g) $6 - 2(4 - x) > 2 + x$
   (h) $3(x - 5) - 4x > 5 - 2x$
   (i) $4 - (x + 7) > 7 - 2(x - 6)$
   (j) $5(2x - 1) - 3(3x + 2) > 0$
2. Choose the greatest number in each set.
   (a) $\{6, {}^-9, 5, {}^-6, {}^-17\}$
   (b) $\{43, 816, {}^-913, {}^-819, 471\}$
   (c) $\{{}^-6, {}^-5, {}^-12, {}^-19, {}^-1\}$
   (d) $\{{}^-1, {}^-2, {}^-3, \ldots\}$

3. True or false? Every positive integer is greater than any negative integer.

4. $5 > 4$ is a true statement. If both sides of the inequality are multiplied by $^-2$, we obtain the statement, $^-2 \cdot 5 > ^-2 \cdot 4$. Is the statement true?

5. $3 < 7$ is a true statement. If both sides of the inequality are multiplied by any negative number, is the statement obtained true?

6. $^-2 > ^-6$ is a true statement. If both sides of the inequality are multiplied by any negative number, is the statement obtained true?

### 3.26   Graphing Ordered Pairs of Integers

In chapter 2, ordered pairs of counting numbers were graphed using a pair of half-lines as shown in Figure 3.4 below. In Figure 3.4, the ordered

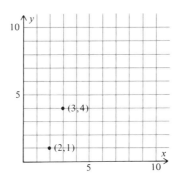

FIGURE 3.4

pairs, $(2, 1)$ and $(3, 4)$, are shown. The point designating $(2, 1)$ is two units to the right of the vertical half-line, and one unit above the horizontal half-line. Similarly, the point designating the ordered pair $(3, 4)$ is three units to the right of the vertical half-line and four units above the horizontal half-line.

To graph ordered pairs of integers such as $(^-4, 5)$, $(^-3, ^-2)$, and $(0, ^-6)$ it is necessary to use a pair of perpendicular lines like those shown in figure 3.5. Nine ordered pairs have been graphed in Figure 3.5. In each case the first number in the ordered pair shows the direction and the distance from the vertical line; the second number in the ordered pair indicates the direction and the distance from the horizontal line.

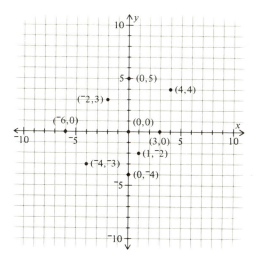

FIGURE 3.5

EXERCISE 3.26.1

1.  In the graph below, find the ordered pair to designate each of the points that are now designated by the capital letters of the alphabet.

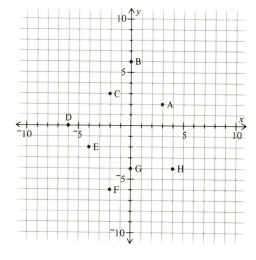

FIGURE 3.6

2. Construct a graph of ordered pairs of integers, and find points to designate the following ordered pairs.

|          |              |          |              |
|----------|--------------|----------|--------------|
| (a) (7, 3) |            | (f) ($^{-}$2, 0) |        |
| (b) (0, $^{-}$3) |      | (g) (0, 0) |               |
| (c) ($^{-}$4, $^{-}$2) || (h) (5, 0) |               |
| (d) (0, 5) |            | (i) ($^{-}$3, 6) |        |
| (e) (4, $^{-}$1) |      | (j) (2, $^{-}$5) |        |

3. If both numbers of an ordered pair are negative integers, where is the point located on the graph?
4. If the first number of an ordered pair is a positive integer and the second number is a negative integer, where is the point located on the graph?
5. If the first number of an ordered pair is zero, where is the point located on the graph?

### 3.27   Properties of the System of Integers

The system of integers is best summarized by listing its properties.

*Addition Properties:*

The Closure Law of Addition: The sum of any two integers is always an integer.

The Commutative Law of Addition: $x + y = y + x$

The Associative Law of Addition: $x + (y + z) = (x + y) + z$

Zero is the Identity Element for Addition: $x + 0 = x$

Every Integer has an Opposite: $x + {}^{-}x = 0$

*Multiplication Properties:*

The Closure Law of Multiplication: The product of any two integers is always an integer.

The Commutative Law of Multiplication: $xy = yx$

The Associative Law of Multiplication: $x(yz) = (xy)z$

One is the Identity Element for Multiplication: $1x = x$

The Multiplication Law of $^{-}1$: $^{-}1x = {}^{-}x$

*A Property Involving both Addition and Multiplication:*

The Distributive Law of Multiplication over Addition:

$$x(y + z) = xy + xz$$

*Properties for Solving Open Sentences:*

(1) Any integer may be added to both members of an equation to generate another equation with exactly the same truth set.

(2) Any integer may be added to both members of an inequality to generate another inequality with exactly the same truth set.

# 4 THE RATIONAL
# NUMBERS

## 4.1 Introduction

In chapter 3, the set of integers, the laws of the set of integers, and some of the applications of these laws for simplifying open expressions and solving equations and inequations were studied.

Using the set of integers as the domain, every equation of the form $x + a = b$, where $a$ and $b$ are integers, has a singleton set as its truth set. For example, $x + 9 = 3$, has a truth set $\{^-6\}$, and the truth set of $x + {}^-8 = {}^-3$ is $\{5\}$.

Some equations of the form $ax = b$, where $a$ and $b$ are integers, do not have an integer that will provide a true statement. Using the set of integers as the domain, the equations $5x = 2$, $^-3x = 13$, and $^-4x = {}^-7$ have the empty set, $\{\ \}$, as their truth set, because in each case there is no integer that can be used to obtain a true statement.

In this chapter another set of numbers, called the rationals, is to be studied. In this chapter we shall see that every equation of the form $ax = b$, where $a$ and $b$ are integers and $a$ is not zero, has a unique rational number that will make the statement true.

### 4.2   The Division of Integers

The set of rational numbers depends upon the division of integers. Problems such as $15 \div 3 = ?$ and $42 \div 7 = ?$ are fairly simple because all of the numbers are counting numbers, but a more comprehensive understanding of division is necessary to correctly answer problems such as $^-21 \div 7 = ?$, $32 \div {}^-8 = ?$, and $^-45 \div {}^-9 = ?$

Every division problem can be written as a multiplication question. $14 \div 2 = ?$ can be written as $2 \cdot ? = 14$. $^-21 \div 7 = ?$ can be written as $7 \cdot ? = {}^-21$.

The division symbol, $\div$, is not commonly used in algebra and higher mathematics. Instead most division problems are written as fractions. For example, $15 \div 3 = ?$ is normally written as $\dfrac{15}{3} = ?$   $32 \div {}^-8 = ?$ is normally written as $\dfrac{32}{^-8} = ?$

The division problem $\dfrac{18}{^-9} = ?$ can be written as $^-9 \cdot ? = 18$. The division problem $\dfrac{^-27}{^-3} = ?$ can be written as $^-3 \cdot ? = {}^-27$.

EXERCISE 4.2.1

Write each division problem as a multiplication question.

1. $\dfrac{20}{4} = ?$

2. $\dfrac{35}{7} = ?$

3. $\dfrac{26}{13} = ?$

4. $\dfrac{19}{19} = ?$

5. $\dfrac{^-10}{5} = ?$

6. $\dfrac{^-30}{6} = ?$

7. $\dfrac{18}{^-3} = ?$

8. $\dfrac{56}{^-7} = ?$

9.  $\dfrac{-24}{-6} = ?$

10.  $\dfrac{-48}{-8} = ?$

11.  $\dfrac{17}{5} = ?$

12.  $\dfrac{-19}{5} = ?$

$\dfrac{15}{5} = ?$ can be written as $5 \cdot ? = 15$. Since $5 \cdot 3 = 15$, then $\dfrac{15}{5}$ must be equal to 3.

$\dfrac{-27}{3} = ?$ can be written as $3 \cdot ? = -27$. Since $3 \cdot -9 = -27$, then $\dfrac{-27}{3}$ must be equal to $-9$.

$\dfrac{24}{-4} = -6$ is a true statement because $-4 \cdot -6 = 24$ is a true statement.

$\dfrac{-8}{3} = 2$ is a false statement because $3 \cdot 2 = -8$ is a false statement.

$\dfrac{-12}{3}$ is equal to the integer $-4$ because $3 \cdot -4 = -12$.

$\dfrac{18}{7}$ is not an integer because there is no integer that can be multiplied by 7 to give a product of 18.

EXERCISE 4.2.2

1.  True or false:

(a)  $\dfrac{30}{-5} = -6$

(b)  $\dfrac{-40}{8} = 5$

(c)  $\dfrac{5}{10} = 2$

(d)  $\dfrac{-12}{-4} = 3$

(e)  $\dfrac{-19}{19} = -1$

(f)  $\dfrac{-50}{10} = -5$

(g) $\dfrac{14}{8} = 2$          (i) $\dfrac{^{-}28}{^{-}14} = 2$

(h) $\dfrac{0}{18} = 0$          (j) $\dfrac{17}{8} = 2$

2. For each of the following find the integer that is equal to the indicated division, or state that there is no such integer.

(a) $\dfrac{28}{4}$          (f) $\dfrac{^{-}5}{7}$

(b) $\dfrac{63}{7}$          (g) $\dfrac{^{-}40}{5}$

(c) $\dfrac{^{-}35}{7}$          (h) $\dfrac{^{-}21}{^{-}21}$

(d) $\dfrac{6}{^{-}1}$          (i) $\dfrac{0}{^{-}7}$

(e) $\dfrac{3}{4}$          (j) $\dfrac{100}{^{-}20}$

$\dfrac{3}{4}$ = ? may be written as $4 \cdot ? = 3$. There is no integer that is equal to

$\dfrac{3}{4}$, but the second equation, $4 \cdot ? = 3$, does show an important property of

$\dfrac{3}{4}$.    $4 \cdot \dfrac{3}{4} = 3$ is a true statement.

$\dfrac{^{-}8}{11}$ = ? may be written as $11 \cdot ? = {}^{-}8$. The question mark must be replaced in both statements by the same number, but that number is not an integer. Its simplest name is $\dfrac{^{-}8}{11}$ and one of its most important properties is shown by the equality $11 \cdot \dfrac{^{-}8}{11} = {}^{-}8$.

$\dfrac{5}{-7}$ is the number that can be multiplied by ⁻7 to produce a product of

5.   $\dfrac{5}{-7}$ ·⁻7 = 5 is a true statement.

$\dfrac{-14}{-3}$ is the number that can be multiplied by ⁻3 to produce a product of

⁻14.   $\dfrac{-14}{-3}$ ·⁻3 = ⁻14 is a true statement.

EXERCISE 4.2.3

What integer replacements for the variables will make the following open sentences true statements?

1.  $\dfrac{6}{11} \cdot 11 = x$

2.  $\dfrac{-14}{27} \cdot 27 = y$

3.  $\dfrac{-8}{16} \cdot 16 = z$

4.  $\dfrac{5}{-9} \cdot x = 5$

5.  $\dfrac{-4}{-19} \cdot y = {}^{-}4$

6.  $\dfrac{z}{7} \cdot 7 = 5$

7.  $\dfrac{8}{x} \cdot {}^{-}3 = 8$

8.  $\dfrac{y}{-13} \cdot {}^{-}13 = {}^{-}4$

9.  $\dfrac{-3}{z} \cdot {}^{-}1 = {}^{-}3$

10.  $\dfrac{-9}{3} \cdot x = {}^{-}9$

## 4.3   The Set of Rational Numbers

Every number of the form $\dfrac{x}{y}$, where $x$ is any integer and $y$ is any integer except zero, is called a rational number.

$\dfrac{3}{4}, \dfrac{-5}{7}, \dfrac{-12}{1}, \dfrac{0}{-6}$, and $\dfrac{-9}{-17}$ are examples of rational numbers because in each case the number above the bar is an integer, and the number below the bar is a non-zero integer. $\frac{7}{0}, \frac{-3}{0}, \frac{0}{0}$, are not rational numbers because in each case the number below the bar is zero, and this is counter to the definition of a rational number.

In the rational number $\frac{-7}{5}$, the integer above the bar, $-7$, is called the numerator, and the integer below the bar, 5, is called the denominator. For the rational number $\frac{15}{-4}$, the integer 15 is the numerator and $-4$ is the denominator.

Every integer can be written as a rational number by using the integer, 1, as the denominator. The integer 3 can be written in the form of a rational number as $\frac{3}{1}$. The integer $-8$ can be written in the form of a rational number as $\frac{-8}{1}$. The integer zero can be written in the form of a rational number as $\frac{0}{1}$.

EXERCISE 4.3.1

1.  Which of the following are rational numbers?

(a) $\frac{16}{5}$    (e) $\frac{5}{0}$

(b) $\frac{-8}{4}$    (f) $\frac{-11}{5}$

(c) $\frac{-11}{16}$    (g) $\frac{-9}{-9}$

(d) $\frac{0}{14}$    (h) $\frac{-17}{0}$

2.  Name the numerator and denominator for each of the following rational numbers:

(a)  $\dfrac{^-8}{2}$

(d)  $\dfrac{14}{^-18}$

(b)  $\dfrac{^-21}{^-6}$

(e)  $\dfrac{0}{^-41}$

(c)  $\dfrac{35}{5}$

3.  Write each in the form of a rational number.
    (a)  16
    (b)  ⁻23
    (c)  4
    (d)  0
    (e)  ⁻1

## 4.4  Equality of Rational Numbers

We have already seen that $\dfrac{8}{2}$ and $\dfrac{4}{1}$ are names for the same rational number. Similarly, $\dfrac{^-35}{5} = \dfrac{^-7}{1}$ and $\dfrac{^-30}{^-5} = \dfrac{6}{1}$ are both true statements. These examples showing that two different numerals may stand for the same rational number are not rare exceptions. In fact, every rational number has an infinite number of numerals. For example, $\dfrac{3}{4}, \dfrac{^-6}{^-8}, \dfrac{^-15}{^-20}, \dfrac{21}{28},$ and $\dfrac{60}{80}$ are all names for the same rational number, and many more numerals for $\dfrac{3}{4}$ could be shown.

There is a relatively simple test that may be applied in testing whether two numerals name the same rational number. The test requires only the multiplication of integers.

To decide whether $\dfrac{3}{4}$ is equal to $\dfrac{^-6}{^-8}$, the truth or falsity of $3 \cdot {}^-8 = 4 \cdot {}^-6$ must be determined. If $3 \cdot {}^-8 = 4 \cdot {}^-6$ is a true statement, then $\dfrac{3}{4} = \dfrac{^-6}{^-8}$ is also a true statement.

To decide whether $\dfrac{-5}{7} = \dfrac{11}{-13}$ is a true or false statement, it must first

be determined whether $-5 \cdot -13 = 7 \cdot 11$ is a true or false statement. Since

$-5 \cdot -13 = 7 \cdot 11$ is a false statement, then $\dfrac{-5}{7} = \dfrac{11}{-13}$ is also a false statement.

*A Test for Equality of Rational Numbers*

To decide whether two rational numbers are equal:

1. Multiply the first numerator and the second denominator.

2. Multiply the first denominator and the second numerator.

3. If the product obtained in step 1 is equal to the product in step 2, then the rational numbers are equal.

EXERCISE 4.4.1

Determine whether each of the following statements of equality are true or false.

1. $\dfrac{2}{5} = \dfrac{4}{10}$

2. $\dfrac{3}{5} = \dfrac{51}{85}$

3. $\dfrac{7}{8} = \dfrac{91}{104}$

4. $\dfrac{5}{12} = \dfrac{9}{20}$

5. $\dfrac{38}{57} = \dfrac{3}{5}$

6. $\dfrac{2}{13} = \dfrac{10}{65}$

7. $\dfrac{3}{12} = \dfrac{18}{72}$

8. $\dfrac{5}{8} = \dfrac{2}{3}$

9. $\dfrac{-3}{5} = \dfrac{3}{-5}$

10. $\dfrac{-4}{7} = \dfrac{4}{7}$

11. $\dfrac{0}{6} = \dfrac{0}{-7}$

12. $\dfrac{4}{-11} = \dfrac{-4}{11}$

13. $\dfrac{-3}{-8} = \dfrac{3}{8}$

14. $\dfrac{0}{15} = \dfrac{0}{49}$

15. $\dfrac{14}{14} = \dfrac{1}{1}$

18. $\dfrac{0}{^{-}1} = \dfrac{0}{23}$

16. $\dfrac{^{-}19}{^{-}19} = \dfrac{1}{1}$

19. $\dfrac{^{-}2}{5} = \dfrac{2}{^{-}5}$

17. $\dfrac{^{-}13}{13} = \dfrac{^{-}1}{1}$

20. $\dfrac{^{-}9}{^{-}16} = \dfrac{9}{16}$

## 4.5   Some Special Cases of Equality

We have already seen that every rational number of the form, $\dfrac{x}{1}$, names an integer. For example, $\dfrac{7}{1}, \dfrac{^{-}6}{1}$, and $\dfrac{0}{1}$ are names for the integers 7, $^{-}$6, and 0. In the remainder of this section some other special cases of equality will be studied.

$\dfrac{0}{41} = \dfrac{0}{^{-}19}$ is a true statement because $0 \cdot {}^{-}19 = 41 \cdot 0$ or $0 = 0$ is a true statement. By the same line of reasoning it can be claimed that any rational number of the form $\dfrac{0}{y}$ is equal to zero. Since zero can be written as $\dfrac{0}{1}$, $\dfrac{0}{y} = \dfrac{0}{1}$ is a true statement because $0 \cdot 1 = y \cdot 0$ is a true statement for any non-zero replacement of $y$. $0 = \dfrac{0}{y}$ is a true statement for any non-zero replacement of $y$.

$\dfrac{12}{12}, \dfrac{^{-}11}{^{-}11}, \dfrac{^{-}45}{^{-}45}$, and $\dfrac{93}{93}$ are all numerals for the integer, 1. $\dfrac{12}{12} = \dfrac{1}{1}$ is a true statement because $12 \cdot 1 = 1 \cdot 12$ is true. $\dfrac{^{-}11}{^{-}11} = \dfrac{1}{1}$ is a true statement because $^{-}11 \cdot 1 = 1 \cdot {}^{-}11$ is true. Any rational number of the form $\dfrac{x}{x}$ where $x \neq 0$ is equal to 1. $\dfrac{x}{x} = \dfrac{1}{1}$ is a true statement because $x \cdot 1 = 1 \cdot x$ is true for any non-zero replacement of the variable, $x$.    $\dfrac{x}{x} = 1$ is a true statement for

any non-zero replacement of $x$.

$\dfrac{^-3}{5} = \dfrac{3}{^-5}$ and $\dfrac{^-4}{^-7} = \dfrac{4}{7}$ are both true statements because $^-3\cdot^-5 = 5\cdot3$

and $^-4\cdot7 = ^-7\cdot4$ are both true statements. Any rational number in the

form of $\dfrac{x}{y}$ is equal to $\dfrac{^-x}{^-y}$, where $^-x$ is read as "the opposite of $x$." Because

$\dfrac{x}{y} = \dfrac{^-x}{^-y}$ always becomes a true statement, it is possible to change the

denominator of any rational number to a positive integer simply by using

the opposite of both integers. $\dfrac{7}{^-10}$ can be changed to the equivalent rational

number, $\dfrac{^-7}{10}$, because $^-7$ is the opposite of 7, and 10 is the opposite of $^-10$.

$\dfrac{^-6}{^-13}$ can be written as $\dfrac{6}{13}$, by changing both $^-6$ and $^-13$ to their opposites.

Any rational number with a negative denominator can be written with a
positive denominator by changing both the numerator and the denominator
to their opposites.

EXERCISE 4.5.1

1. What integer is named by each of the following rational
numbers?

(a) $\dfrac{12}{1}$                    (d) $\dfrac{15}{15}$

(b) $\dfrac{^-14}{1}$                    (e) $\dfrac{^-12}{3}$

(c) $\dfrac{0}{^-3}$                    (f) $\dfrac{0}{49}$

(g) $\dfrac{^-16}{^-16}$   (i) $\dfrac{0}{^-25}$

(h) $\dfrac{812}{812}$   (j) $\dfrac{^-93}{^-93}$

2.  Write each rational number with a positive denominator.

(a) $\dfrac{5}{^-9}$   (f) $\dfrac{14}{^-1}$

(b) $\dfrac{^-12}{^-7}$   (g) $\dfrac{^-19}{^-7}$

(c) $\dfrac{^-8}{^-15}$   (h) $\dfrac{2}{^-17}$

(d) $\dfrac{3}{^-13}$   (i) $\dfrac{803}{^-41}$

(e) $\dfrac{^-47}{^-8}$   (j) $\dfrac{^-19}{^-73}$

## 4.6   Multiplication of Rational Numbers

The multiplication of rational numbers is described in a manner that retains the use of the multiplication of integers as previously studied. For example, $2 \cdot {}^-3 = {}^-6$ is a true statement for the integers and, therefore,

$\dfrac{2}{1} \cdot \dfrac{^-3}{1} = \dfrac{^-6}{1}$ must be a true statement for the multiplication of rational

numbers. Although this example seems trivial, the student who remembers it will find it an easy reminder should he later question this method of multiplying the rational numbers.

*Describing the Multiplication of Rational Numbers*

To multiply any two rational numbers.

1.  The numerators are multiplied to give the numerator of the product.

2.  The denominators are multiplied to give the denominator of the product.

As examples of multiplication of rational numbers, study the following:

$$\frac{3}{5} \cdot \frac{^-4}{7} = \frac{3 \cdot {}^-4}{5 \cdot 7} = \frac{^-12}{35}$$

$$\frac{^-4}{9} \cdot \frac{^-11}{3} = \frac{^-4 \cdot {}^-11}{9 \cdot 3} = \frac{44}{27}$$

$$\frac{^-6}{7} \cdot 2 = \frac{^-6}{7} \cdot \frac{2}{1} = \frac{^-6 \cdot 2}{7 \cdot 1} = \frac{^-12}{7}$$

$$\frac{13}{9} \cdot 0 = \frac{13}{9} \cdot \frac{0}{1} = \frac{13 \cdot 0}{9 \cdot 1} = \frac{0}{9} = 0$$

EXERCISE 4.6.1

Find the product:

1. $\dfrac{5}{8} \cdot \dfrac{^-3}{7}$

2. $\dfrac{^-4}{5} \cdot \dfrac{^-2}{3}$

3. $\dfrac{^-8}{9} \cdot 5$

4. $\dfrac{4}{11} \cdot {}^-3$

5. $\dfrac{19}{5} \cdot \dfrac{^-1}{2}$

6. $\dfrac{27}{7} \cdot \dfrac{^-2}{5}$

7. $\dfrac{^-16}{3} \cdot 0$

8. $\dfrac{47}{5} \cdot 1$

9. $6 \cdot \dfrac{^-2}{11}$

10. $^-4 \cdot \dfrac{^-6}{17}$

## 4.7  Simplifying Rational Numbers

We have seen earlier that $\dfrac{^-7}{9}$, $\dfrac{21}{^-27}$, $\dfrac{^-35}{45}$, and $\dfrac{^-70}{90}$ are all names for

the same rational number. We say that $\dfrac{^-7}{9}$ is the simplest form of the name for the number, and in this section a method for simplifying rational numbers is studied.

An intermediate concept called highest common factors is required to properly understand the simplification of rational numbers. 2 is a factor of 6 because $2 \cdot 3 = 6$. 7 is a factor of $^-35$ because $7 \cdot ^-5 = ^-35$. 1 is a factor of 8 because $1 \cdot 8 = 8$.

The set of all positive factors of 10 is $\{1, 2, 5, 10\}$ because each of the elements of the set can be evenly divided into 10.

The set of all positive factors of $^-28$ is $\{1, 2, 4, 7, 14, 28\}$ because each of the elements of the set can be evenly divided into $^-28$.

EXERCISE 4.7.1

Find the set of all positive factors of each of the following integers.

1. 6
2. 15
3. 32
4. $^-20$
5. $^-30$
6. 72
7. What number is a positive factor of all integers?

The rational number $\dfrac{^-10}{14}$ can be simplified because $^-10$ and 14 each have a factor of 2. The simplification of $\dfrac{^-10}{14}$ is accomplished in the following steps:

$$\frac{^-10}{14} = \frac{2 \cdot ^-5}{2 \cdot 7} = \frac{2}{2} \cdot \frac{^-5}{7} = 1 \cdot \frac{^-5}{7} = \frac{^-5}{7}$$

Notice that the simplification of $\dfrac{^-10}{14}$ is dependent upon the multiplication of rational numbers and the fact that $\dfrac{2}{2}$ is a numeral for positive 1.

The rational expression $\dfrac{^-9xy}{15xz}$ is simplified by use of the following steps:

$$\frac{^-9xy}{15xz} = \frac{3}{3} \cdot \frac{^-3}{5} \cdot \frac{x}{x} \cdot \frac{y}{z} = 1 \cdot \frac{^-3}{5} \cdot 1 \cdot \frac{y}{z} = \frac{^-3y}{5z}$$

The simplification of $\dfrac{-9xy}{15xz}$ is dependent upon the fact that $\dfrac{3}{3}$ and $\dfrac{x}{x}$ are numerals for positive 1.

A rational number is completely simplified when:

1. The only common positive factor of the numerator and denominator is positive one.

2. The denominator is positive.

EXERCISE 4.7.2

Simplify:

1. $\dfrac{-18}{24}$                9. $\dfrac{10x}{7x}$

2. $\dfrac{-12}{-16}$               10. $\dfrac{5x}{10x}$

3. $\dfrac{14}{-49}$               11. $\dfrac{-13x}{-3x}$

4. $\dfrac{60}{32}$               12. $\dfrac{4xy}{10y}$

5. $\dfrac{-19}{-25}$               13. $\dfrac{-10xyz}{5xy}$

6. $\dfrac{36}{-27}$               14. $\dfrac{-16xz}{-6yz}$

7. $\dfrac{-40}{55}$               15. $\dfrac{14x}{-8xy}$

8. $\dfrac{-63}{-99}$               16. $\dfrac{-3xyz}{-3xyz}$

17. Explain the assertion: The simplification of rational numbers is dependent upon the Multiplication Law of One.

## 4.8 Simplifying Multiplication Expressions

To multiply two rational numbers such as $\dfrac{5}{4} \cdot \dfrac{7}{3}$, the numerators are multiplied to find the numerator of the product, and the denominators are multiplied to find the denominator of the product. Hence,

$$\frac{5}{4} \cdot \frac{7}{3} = \frac{5 \cdot 7}{4 \cdot 3} = \frac{35}{12}.$$

Sometimes the multiplication of rational numbers can be simplified because there are common factors in the numerators and denominators. For example, $\dfrac{60}{80} \cdot \dfrac{5}{7}$, is more easily accomplished as $\dfrac{3}{4} \cdot \dfrac{5}{7}$, because $\dfrac{60}{80}$ is equal to $\dfrac{3}{4}$.

As another example consider the problem, $\dfrac{2}{5} \cdot \dfrac{15}{7}$. To simplify the computation of this problem the denominators can be interchanged, because of the Commutative Law of Multiplication, and the multiplication completed as follows:

$$\frac{2}{5} \cdot \frac{15}{7} = \frac{2}{7} \cdot \frac{15}{5} = \frac{2}{7} \cdot \frac{3}{1} = \frac{6}{7}$$

Notice that the previous example is dependent upon the simplification of $\dfrac{15}{5}$ to $\dfrac{3}{1}$.

To simplify $\dfrac{8x}{5y} \cdot \dfrac{3y}{16z}$, the following steps are used:

$$\frac{8x}{5y} \cdot \frac{3y}{16z} = \frac{8x}{16z} \cdot \frac{3y}{5y} = \frac{x}{2z} \cdot \frac{3}{5} = \frac{3x}{10z}$$

The simplification is possible because $\dfrac{8x}{16z}$ can be simplified to $\dfrac{x}{2z}$ and $\dfrac{3y}{5y}$ can be simplified to $\dfrac{3}{5}$.

EXERCISE 4.8.1

Multiply (first simplify wherever possible):

1. $\dfrac{6}{7} \cdot \dfrac{5}{25}$

11. $\dfrac{-4}{9} \cdot \dfrac{-15}{22}$

2. $\dfrac{-3}{5} \cdot \dfrac{5}{6}$

12. $\dfrac{-2}{5} \cdot \dfrac{-15}{7}$

3. $\dfrac{14}{8} \cdot \dfrac{-3}{5}$

13. $\dfrac{4x}{3y} \cdot \dfrac{-9}{x}$

4. $\dfrac{-7}{2} \cdot \dfrac{-6}{11}$

14. $\dfrac{-3x}{5} \cdot \dfrac{10}{x}$

5. $\dfrac{-8}{7} \cdot \dfrac{-3}{4}$

15. $\dfrac{6x}{9y} \cdot \dfrac{2z}{5x}$

6. $\dfrac{5}{9} \cdot \dfrac{-12}{25}$

16. $\dfrac{-x}{3y} \cdot \dfrac{-2y}{7x}$

7. $\dfrac{5}{12} \cdot \dfrac{16}{9}$

17. $\dfrac{-4z}{5z} \cdot \dfrac{15x}{8}$

8. $\dfrac{-2}{9} \cdot \dfrac{-12}{13}$

18. $\dfrac{5x}{7y} \cdot \dfrac{-14y}{25x}$

9. $\dfrac{-6}{5} \cdot \dfrac{10}{21}$

19. $\dfrac{4x}{7y} \cdot \dfrac{7y}{4x}$

10. $\dfrac{-4}{7} \cdot \dfrac{21}{10}$

20. $\dfrac{-4x}{9y} \cdot \dfrac{-9y}{4x}$

## 4.9   Addition of Rational Numbers

To add $\dfrac{5}{7}$ and $\dfrac{3}{7}$, the numerators are added, and the denominator is

left unchanged, as follows: $\dfrac{5}{7} + \dfrac{3}{7} = \dfrac{5+3}{7} = \dfrac{8}{7}$.

To add $\dfrac{4}{5}$ and $\dfrac{2}{3}$, another method is used because the denominators of the two rational numbers are not the same. The steps to complete the addition problem, $\dfrac{4}{5} + \dfrac{2}{3}$, are shown below:

$$\frac{4}{5} + \frac{2}{3} = \frac{4 \cdot 3 + 5 \cdot 2}{5 \cdot 3} = \frac{12 + 10}{15} = \frac{22}{15}$$

Notice that the previous example was accomplished by using the following rule for the addition of rational numbers.

*Describing the Addition of Rational Numbers*

1.  To find the numerator of the sum:

    (a)  Multiply the first numerator by the second denominator.
    (b)  Multiply the first denominator by the second numerator.
    (c)  Add the products obtained in steps (a) and (b).

2.  To find the denominator of the sum, multiply the two denominators.

The procedure outlined above will *always* result in the correct addition of two rational numbers, because the method assures a common denominator. This may not be apparent from the description given above, and this method does not always give the sum in the simplest form, but it does give a simple procedure for finding the sum of any two rational numbers. Below are some examples following the procedure described above:

$$\frac{6}{11} + \frac{4}{7} = \frac{6 \cdot 7 + 11 \cdot 4}{11 \cdot 7} = \frac{42 + 44}{77} = \frac{86}{77}$$

$$\frac{3}{4} + \frac{5}{8} = \frac{3 \cdot 8 + 4 \cdot 5}{4 \cdot 8} = \frac{24 + 20}{32} = \frac{44}{32} = \frac{11}{8}$$

$$\frac{3}{7} + \frac{2}{7} = \frac{3 \cdot 7 + 7 \cdot 2}{7 \cdot 7} = \frac{21 + 14}{49} = \frac{35}{49} = \frac{5}{7}$$

$$\frac{6}{x} + \frac{5}{y} = \frac{6 \cdot y + x \cdot 5}{xy} = \frac{6y + 5x}{xy}$$

Among the four examples shown above, the second and third show situations which may be more easily added without relying upon the procedure given for the addition of rational numbers. Notice, however, that the procedure is capable of supplying the correct answer for any addition problem in a straightforward method which makes it an important skill for dealing with rational expressions.

EXERCISE 4.9.1

Add:

1. $\dfrac{1}{5} + \dfrac{2}{7}$     11. $\dfrac{3}{4} + \dfrac{5}{6}$

2. $\dfrac{5}{9} + \dfrac{3}{8}$     12. $\dfrac{5}{8} + \dfrac{1}{10}$

3. $\dfrac{4}{3} + \dfrac{2}{5}$     13. $\dfrac{^-3}{5} + \dfrac{7}{15}$

4. $\dfrac{3}{4} + \dfrac{1}{5}$     14. $\dfrac{8}{9} + \dfrac{^-7}{12}$

5. $\dfrac{4}{3} + \dfrac{1}{7}$     15. $\dfrac{^-3}{8} + 4$

6. $\dfrac{6}{11} + \dfrac{2}{1}$     16. $\dfrac{5}{x} + \dfrac{2}{y}$

7. $\dfrac{^-4}{5} + \dfrac{3}{2}$     17. $\dfrac{3}{x} + \dfrac{^-5}{y}$

8. $\dfrac{^-2}{7} + \dfrac{^-2}{3}$     18. $\dfrac{x}{2} + \dfrac{y}{3}$

9. $\dfrac{4}{5} + \dfrac{^-6}{11}$     19. $\dfrac{2x}{3y} + \dfrac{1}{5}$

10. $^-6 + \dfrac{3}{4}$     20. $\dfrac{x}{y} + \dfrac{z}{w}$

## 4.10  Opposites of Rational Numbers

One of the most important properties of the set of integers is the fact that every integer has an opposite. $^-2$ is the opposite of 2 because they have a sum of zero, $2 + {}^-2 = 0$. 11 is the opposite of $^-11$, because $^-11 + 11 = 0$.

Since the numerator of every rational number is an integer, and every integer has an opposite, every rational number will also have an opposite.

For example, to find the opposite of $\dfrac{3}{5}$ we select $\dfrac{^-3}{5}$, because 3 and $^-3$ are opposites.

$$\frac{3}{5} + \frac{^-3}{5} = \frac{3 + {}^-3}{5} = \frac{0}{5} = 0$$

To find the opposite of $\dfrac{^-9}{4}$, we select $\dfrac{9}{4}$, because $^-9$ and 9 are opposites.

$$\frac{^-9}{4} + \frac{9}{4} = \frac{^-9 + 9}{4} = \frac{0}{4} = 0$$

EXERCISE 4.10.1

Find the opposite of each of the following rational numbers. Check your answers by showing that the number and the opposite have a sum of zero.

1. $\dfrac{7}{13}$

2. $\dfrac{^-11}{5}$

3. $\dfrac{^-7}{6}$

4. $\dfrac{19}{14}$

5. $\dfrac{0}{7}$

6. $\dfrac{5}{x}$

7. $\dfrac{^-3x}{y}$

8. $\dfrac{5xy}{z}$

9. $\dfrac{4}{x + y}$

10. $\dfrac{x}{y}$

## 4.11   Using the Minus Sign with Rational Numbers

$\dfrac{5}{7} - \dfrac{3}{4}$ is an addition problem in which the minus sign indicates that the opposite of $\dfrac{3}{4}$ is to be added to $\dfrac{5}{7}$. $\dfrac{5}{7} - \dfrac{3}{4}$ is evaluated in the following steps:

$$\frac{5}{7} - \frac{3}{4} = \frac{5}{7} + \frac{^-3}{4} = \frac{20 + ^-21}{28} = \frac{^-1}{28}$$

Whenever the minus sign appears between two rational numbers, it indicates that the opposite of the second number is to be added to the first. Two more examples of the use of the minus sign are shown below:

$$\frac{^-5}{6} - \frac{^-2}{5} = \frac{^-5}{6} + \frac{2}{5} = \frac{^-25 + 12}{30} = \frac{^-13}{30}$$

$$\frac{y}{3} - \frac{x}{4} = \frac{y}{3} + \frac{^-x}{4} = \frac{4y + ^-3x}{12} = \frac{4y - 3x}{12}$$

EXERCISE 4.11.1

Evaluate:

1. $\dfrac{4}{7} - \dfrac{2}{3}$

2. $\dfrac{5}{6} - \dfrac{^-3}{8}$

3. $\dfrac{^-4}{3} - \dfrac{1}{2}$

4. $\dfrac{^-6}{5} - \dfrac{^-3}{2}$

5. $\dfrac{4}{9} - \dfrac{1}{3}$

6. $\dfrac{7}{12} - \dfrac{^-3}{8}$

7. $\dfrac{x}{5} - \dfrac{y}{7}$

8. $\dfrac{2}{x} - \dfrac{^-3}{y}$

9. $\dfrac{^-5x}{2} - \dfrac{3x}{5}$

10. $\dfrac{9x}{4yz} - \dfrac{9x}{4yz}$

## 4.12   Use of Parentheses in Addition Expressions

Addition may be performed on only two numbers at one time. To add three numbers or more, parentheses (or square brackets) may be used to show the groupings of two rational numbers to indicate the order in which the numbers are to be added. The steps involved in evaluating $\left(\dfrac{5}{6} - \dfrac{2}{5}\right) + \dfrac{7}{2}$ are shown below.

$$\left(\frac{5}{6} - \frac{2}{5}\right) + \frac{7}{2} = \left(\frac{5}{6} + \frac{^{-}2}{5}\right) + \frac{7}{2}$$

$$= \frac{25 + {^{-}12}}{30} + \frac{7}{2}$$

$$= \frac{13}{30} + \frac{7}{2}$$

$$= \frac{26 + 210}{60}$$

$$= \frac{236}{60} \quad \text{or} \quad \frac{59}{15}$$

Notice the method used in handling the minus sign in the above example, and again the way it is used in the following example. To simplify $\dfrac{3}{x} - \left(\dfrac{4}{5} - \dfrac{1}{2}\right)$, the following steps are used:

$$\frac{3}{x} - \left(\frac{4}{5} - \frac{1}{2}\right) = \frac{3}{x} - \left(\frac{4}{5} + \frac{^{-}1}{2}\right)$$

$$= \frac{3}{x} - \left(\frac{8 + {^{-}5}}{10}\right)$$

$$= \frac{3}{x} - \frac{3}{10}$$

$$= \frac{3}{x} + \frac{^{-}3}{10}$$

$$= \frac{30 + {^{-}3x}}{x \cdot 10}$$

$$= \frac{30 - 3x}{10x}$$

EXERCISE 4.12.1

Evaluate:

1. $\left(\dfrac{1}{3} + \dfrac{^{-}1}{2}\right) + \dfrac{2}{5}$

2. $\left(\dfrac{2}{3} - \dfrac{1}{4}\right) - \dfrac{1}{5}$

3. $\left(\dfrac{^{-}1}{8} + \dfrac{1}{3}\right) - \dfrac{^{-}3}{4}$

4. $\left(\dfrac{2}{7} + 2\right) - \dfrac{1}{5}$

5. $\dfrac{1}{3} + \left(\dfrac{^{-}1}{2} + \dfrac{2}{5}\right)$

6. $\dfrac{2}{3} - \left(\dfrac{1}{4} + \dfrac{1}{5}\right)$

7. $\dfrac{^{-}1}{8} + \left(\dfrac{1}{3} - \dfrac{^{-}3}{4}\right)$

8. $\dfrac{2}{7} + \left(2 - \dfrac{1}{5}\right)$

9. $\left(\dfrac{3}{5} - \dfrac{2}{3}\right) - \dfrac{x}{2}$

10. $\dfrac{4}{x} - \left(\dfrac{1}{2} - 2\right)$

## 4.13  Laws of Addition for the Rational Numbers

There are four laws of addition for the rational numbers. They are similar to the laws of addition for the system of integers.

*The Commutative Law of Addition*

The sum of two rational numbers is the same regardless of the order in which they are added. $x + y = y + x$ will become a true statement for any replacements of $x$ and $y$ by rational numbers.

According to the Commutative Law of Addition each of the following open sentences becomes a true statement for any replacements of the variables.

$$\frac{5}{3} + \frac{x}{7} = \frac{x}{7} + \frac{5}{3}$$

$$\frac{4}{17} - \frac{6}{x} = \frac{^{-}6}{x} + \frac{4}{17}$$

$$x + \frac{7}{2} = \frac{7}{2} + x$$

$$\frac{^-4}{3} + y = y - \frac{4}{3}$$

$$\left(\frac{5}{3} + x\right) + \frac{^-5}{3} = \frac{^-5}{3} + \left(\frac{5}{3} + x\right)$$

$$\frac{2}{3} + \left(y - \frac{2}{3}\right) = \left(y - \frac{2}{3}\right) + \frac{2}{3}$$

## The Associative Law of Addition

The addition of three rational numbers is the same whether the sum of the first two is added to the third or the first is added to the sum of the last two. $(x + y) + z = x + (y + z)$ will become a true statement for any replacements of $x$, $y$, and $z$ by rational numbers.

According to the Associative Law of Addition each of the following open sentences becomes a true statement for any rational number replacements of the variables.

$$\left(x + \frac{2}{3}\right) - \frac{5}{7} = x + \left(\frac{2}{3} - \frac{5}{7}\right)$$

$$\left(y - \frac{3}{4}\right) - \frac{7}{5} = y + \left(\frac{^-3}{4} - \frac{7}{5}\right)$$

$$\left(\frac{2}{5} - z\right) + y = \frac{2}{5} + (^-z + y)$$

$$\frac{4}{7} - \left(x + \frac{4}{9}\right) = \left(\frac{4}{7} - x\right) + \frac{^-4}{9}$$

## Zero is the Identity Element for Addition

The sum of any rational number and zero is that same rational number. $x + 0 = x$ becomes a true statement for any rational number replacement of the variable.

Each of the following open sentences becomes a true statement for any rational number replacements of the variables because zero is the identity element for addition.

$$x + \frac{0}{19} = x$$

$$x + \left(\frac{4}{3} + \frac{^-4}{3}\right) = x$$

$$x + \left(\frac{y}{5} + \frac{^-y}{5}\right) = x$$

$$\frac{2x}{9} + \frac{0}{11} = \frac{2x}{9}$$

*Every Rational Number has an Opposite*

Given any first rational number, there is another rational number such that their sum is zero.

Each of the following open sentences becomes a true statement for any replacement of the variables because the sum of any rational number and its opposite is zero.

$$\frac{x}{4} + \frac{^-x}{4} = 0$$

$$\frac{5xy}{z} + \frac{^-5xy}{z} = 0$$

$$\frac{^-2x}{y} + \frac{^-(^-2x)}{y} = 0$$

$$\frac{4 + x}{3y} + \frac{^-(4 + x)}{3y} = 0$$

EXERCISE 4.13.1

For each open sentence, state the Law of Addition which claims that it must become a true statement for any replacements of the variables.

1.  $\frac{2}{5} + x = x + \frac{2}{5}$

2. $\dfrac{x}{4} + \dfrac{0}{9} = \dfrac{x}{4}$

3. $\dfrac{7}{3x} + \dfrac{^-7}{3x} = 0$

4. $\left(\dfrac{5}{3} - x\right) + \dfrac{2}{3} = \dfrac{5}{3} + \left(^-x + \dfrac{2}{3}\right)$

5. $\dfrac{x}{5} - 4 = {}^-4 + \dfrac{x}{5}$

6. $5 + \left(\dfrac{3}{5} - x\right) = \left(5 + \dfrac{3}{5}\right) - x$

7. $\dfrac{5x}{9} + \left(\dfrac{4}{7} - \dfrac{4}{7}\right) = \dfrac{5x}{9}$

8. $\dfrac{9}{4x} + \dfrac{^-9}{4x} = 0$

9. $\left(x - \dfrac{1}{3}\right) + \dfrac{2}{5} = \dfrac{2}{5} + \left(x - \dfrac{1}{3}\right)$

10. $\dfrac{x - 4}{6} + \dfrac{^-(x - 4)}{6} = 0$

11. $9 + \left(\dfrac{8}{y} - \dfrac{8}{y}\right) = 9$

12. $\dfrac{4x}{7} - \dfrac{3y}{5} = \dfrac{^-3y}{5} + \dfrac{4x}{7}$

13. $\left(x - \dfrac{3}{5}\right) + \dfrac{4}{7} = x + \left(\dfrac{^-3}{5} + \dfrac{4}{7}\right)$

14. $\dfrac{2}{5 - 3x} + \dfrac{^-2}{5 - 3x} = 0$

15. $8 + \dfrac{0}{9y - 4x} = 8$

16. $9 + \left(x - \dfrac{3}{4}\right) = \left(x - \dfrac{3}{4}\right) + 9$

17. $\dfrac{2}{5} - \dfrac{x}{7} = \dfrac{^-x}{7} + \dfrac{2}{5}$

18. $x + \left(\dfrac{^-6}{5} - 3\right) = \left(x - \dfrac{6}{5}\right) - 3$

19. $\dfrac{2x}{3} - \left(\dfrac{x}{5} + 2\right) = \left(\dfrac{2x}{3} - \dfrac{x}{5}\right) - 2$

20. $\left(\dfrac{2}{3} + \dfrac{x}{5}\right) - \dfrac{8x}{3} = \dfrac{2}{3} + \left(\dfrac{x}{5} - \dfrac{8x}{3}\right)$

### 4.14  Simplifying Addition Expressions

The four laws of addition used to simplify open expressions, such as $\left(\dfrac{3}{4} + x\right) + \dfrac{2}{3}$ and $\left(\dfrac{2x}{3} - \dfrac{1}{5}\right) + \left(\dfrac{3}{8} - \dfrac{2x}{3}\right)$ are the Commutative Law of Addition, the Associative Law of Addition, the Identity Element for Addition, and the Law of Opposites for Addition.

To simplify $\left(\dfrac{3}{4} + x\right) + \dfrac{2}{3}$, the following steps are used:

$$\left(\dfrac{3}{4} + x\right) + \dfrac{2}{3} = \left(x + \dfrac{3}{4}\right) + \dfrac{2}{3}$$

$$= x + \left(\dfrac{3}{4} + \dfrac{2}{3}\right)$$

$$= x + \dfrac{17}{12}$$

To simplify $\left(\dfrac{2x}{3} - \dfrac{1}{5}\right) + \left(\dfrac{3}{8} - \dfrac{2x}{3}\right)$, the following steps are used:

$$\left(\frac{2x}{3} - \frac{1}{5}\right) + \left(\frac{3}{8} - \frac{2x}{3}\right) = \left(\frac{2x}{3} + \frac{^{-1}}{5}\right) + \left(\frac{3}{8} + \frac{^{-2x}}{3}\right)$$

$$= \left(\frac{^{-1}}{5} + \frac{3}{8}\right) + \left(\frac{2x}{3} + \frac{^{-2x}}{3}\right)$$

$$= \frac{7}{40} + 0$$

$$= \frac{7}{40}$$

EXERCISE 4.14.1

Simplify the following open expressions:

1. $\left(x - \dfrac{1}{3}\right) + \dfrac{1}{2}$

2. $\left(y + \dfrac{4}{7}\right) - \dfrac{1}{8}$

3. $\left(\dfrac{3}{4} + z\right) + \dfrac{1}{5}$

4. $\left(\dfrac{5}{6} - x\right) - \dfrac{1}{3}$

5. $\left(\dfrac{5y}{4} - 2\right) + \dfrac{8}{5}$

6. $\left(\dfrac{2x}{5} + \dfrac{6}{5}\right) - \dfrac{1}{2}$

7. $\left(5 - \dfrac{8x}{3}\right) + \left(\dfrac{^{-11}}{4} + \dfrac{8x}{3}\right)$

8. $\left(\dfrac{9}{5} - \dfrac{5x}{8}\right) + \dfrac{5x}{8}$

9. $\left(\dfrac{3x}{7} - \dfrac{12}{25}\right) + \dfrac{12}{25}$

10. $\left(\dfrac{8}{15} - \dfrac{3x}{2}\right) + \dfrac{3x}{2}$

11. $\left(\dfrac{2x}{5} - \dfrac{1}{3}\right) + \left(\dfrac{5}{6} - \dfrac{2x}{5}\right)$

12. $\left(\dfrac{4}{7} - \dfrac{16x}{3}\right) + \left(\dfrac{3}{5} - \dfrac{1}{2}\right)$

13. $\left(\dfrac{6x}{7} - \dfrac{3}{4}\right) + \left(\dfrac{5}{4} - \dfrac{6x}{7}\right)$

14. $\left(\dfrac{3x}{2} + 5\right) + \left(\dfrac{6x}{7} - \dfrac{3x}{2}\right)$

15. $\left(\dfrac{7x}{8} - \dfrac{1}{3}\right) + \dfrac{5}{6}$     18. $\left(2 - \dfrac{5}{9}\right) + \left(\dfrac{4x}{3} - \dfrac{3}{5}\right)$

16. $\left(\dfrac{^-3x}{8} + \dfrac{5}{3}\right) + \left(\dfrac{^-7}{8} + \dfrac{3x}{8}\right)$     19. $\left(5 - \dfrac{x}{7}\right) + \left(\dfrac{1}{3} - \dfrac{2}{7}\right)$

17. $\left(3 - \dfrac{5x}{7}\right) + \left(\dfrac{7}{4} - \dfrac{2}{5}\right)$     20. $\left(7 + \dfrac{17x}{3}\right) + \left(\dfrac{^-43}{8} - \dfrac{17x}{3}\right)$

## 4.15   Multiplication Laws for the Rational Numbers

There are four laws of multiplication for the set of rational numbers. Three of these laws are similar to the laws for the system of integers; the fourth law has no counterpart in the system of integers.

*The Commutative Law of Multiplication*

The product of two rational numbers is the same regardless of the order in which they are multiplied. $xy = yx$ becomes a true statement for any rational number replacements of the variables, $x$ and $y$.

According to the Commutative Law of Multiplication, each of the following open sentences becomes a true statement for any replacements of the variables.

$$\frac{3}{4}x = x \cdot \frac{3}{4}$$

$$\left(\frac{5}{6}x\right) \cdot \frac{^-1}{3} = \frac{^-1}{3} \cdot \left(\frac{5}{6}x\right)$$

$$\left(\frac{5}{7}x - \frac{6}{13}\right) \cdot \frac{^-8}{5} = \frac{^-8}{5} \cdot \left(\frac{5}{7}x - \frac{6}{13}\right)$$

*The Associative Law of Multiplication*

The multiplication of three rational numbers is the same whether the product of the first two is multiplied by the third, or the first is multiplied by the product of the last two. $(xy)z = x(yz)$ becomes a true statement for any rational number replacements of the variables, $x$, $y$, and $z$.

According to the Associative Law of Multiplication each of the following open sentences becomes a true statement for any replacements of the variables, $x$ and $y$.

$$\frac{^-4}{5} \cdot \left( \frac{6}{7} x \right) = \left( \frac{^-4}{5} \cdot \frac{6}{7} \right) x$$

$$\frac{^-4}{3} \cdot \left( \frac{^-8}{9} y \right) = \left( \frac{^-4}{3} \cdot \frac{^-8}{9} \right) y$$

$$\left( \frac{2}{3} x \right) \cdot \frac{^-6}{11} = \frac{2}{3} \left( x \cdot \frac{^-6}{11} \right)$$

*One is the Identity Element for Multiplication*

The product of any rational number and positive one is that same rational number. $1x = x$ becomes a true statement for any rational number replacement of the variable, $x$.

Each of the following open sentences becomes a true statement for any replacements of the variable because one is the identity element for multiplication.

$$1z = z$$

$$1 \left( \frac{^-3}{4} xy \right) = \frac{^-3}{4} xy$$

$$1 \left( \frac{^-2x}{9} + \frac{13}{4} \right) = \left( \frac{^-2x}{9} + \frac{13}{4} \right)$$

*Every Rational Number except Zero Has a Reciprocal*

Given any rational number except zero, there is another rational number such that their product is positive one. $x \cdot \dfrac{1}{x} = 1$, for all rational numbers, $x$, $x \neq 0$.

Each of the following is an example of the claim that every rational number has a reciprocal.

$\dfrac{-5}{7}$ is the reciprocal of $\dfrac{-7}{5}$, because $\dfrac{-5}{7} \cdot \dfrac{-7}{5} = 1$

$\dfrac{8}{3}$ is the reciprocal of $\dfrac{3}{8}$, because $\dfrac{8}{3} \cdot \dfrac{3}{8} = 1$

$\dfrac{-5}{x}$ is the reciprocal of $\dfrac{-x}{5}$, because $\dfrac{-5}{x} \cdot \dfrac{-x}{5} = 1$

$\dfrac{y}{11}$ is the reciprocal of $\dfrac{11}{y}$, because $\dfrac{y}{11} \cdot \dfrac{11}{y} = 1$

EXERCISE 4.15.1

1.  Find the reciprocal of each of the following rational numbers.

    (a) $\dfrac{5}{13}$          (f) $\dfrac{-13}{1}$

    (b) $\dfrac{7}{2}$          (g) $5$

    (c) $\dfrac{-6}{19}$          (h) $-21$

    (d) $\dfrac{-5}{3}$          (i) $\dfrac{-7}{10}$

    (e) $\dfrac{8}{1}$          (j) $\dfrac{14}{23}$

2.  For each open sentence, state the Multiplication Law that claims it will become a true statement for any replacements of the variables.

    (a) $\dfrac{-13}{5} z = z \cdot \dfrac{-13}{5}$

    (b) $\dfrac{4}{3}\left(\dfrac{-8}{57} y\right) = \left(\dfrac{4}{3} \cdot \dfrac{-8}{57}\right) y$

(c) $1\left(\dfrac{-19}{6}\,w\right) = \dfrac{-19}{6}\,w$

(d) $\dfrac{-5}{7z} \cdot \dfrac{-7z}{5} = 1$

(e) $\left(\dfrac{23}{12} - \dfrac{6x}{5}\right) = 1\left(\dfrac{23}{12} - \dfrac{6x}{5}\right)$

(f) $\left(\dfrac{33}{47}\,x\right)y = \dfrac{33}{47}\,(xy)$

(g) $\left(\dfrac{-73}{19}\,z\right) \cdot \dfrac{-7}{3} = \dfrac{-7}{3} \cdot \left(\dfrac{-73}{19}\,z\right)$

(h) $\dfrac{2x}{3} \cdot \dfrac{3}{2x} = 1$

(i) $\dfrac{4}{7}\left(\dfrac{-6}{19}\,z + \dfrac{1}{4}\right) = \left(\dfrac{-6}{19}\,z + \dfrac{1}{4}\right) \cdot \dfrac{4}{7}$

(j) $\dfrac{7}{5} \cdot \left(\dfrac{-9}{23}\,y\right) = \left(\dfrac{7}{5} \cdot \dfrac{-9}{23}\right)y$

(k) $\dfrac{-15}{11}\,xyz = 1\left(\dfrac{-15}{11}\,xyz\right)$

(l) $\left(\dfrac{4}{91} \cdot \dfrac{-61}{9}\right)w = \dfrac{4}{91}\left(\dfrac{-61}{9}\,w\right)$

## 4.16   Simplifying Multiplication Expressions

Using the Commutative and Associative Laws of Multiplication, a number of open expressions can be simplified. For example, $5\left(\dfrac{3}{4}\,x\right)$ is simplified as follows:

$$5\left(\dfrac{3}{4}\,x\right) = \left(5 \cdot \dfrac{3}{4}\right)x$$

$$= \dfrac{15}{4}\,x$$

and $\left(\dfrac{-2}{3}y\right) \cdot \dfrac{4}{5}$ is simplified as follows:

$$\left(\frac{-2}{3}y\right) \cdot \frac{4}{5} = \frac{4}{5} \cdot \left(\frac{-2}{3}y\right)$$

$$= \left(\frac{4}{5} \cdot \frac{-2}{3}\right)y$$

$$= \frac{-8}{15}y$$

The simplification of $^-y\left(\dfrac{-2}{3}x\right)$ is dependent upon the fact that $^-y$ is equivalent to $^-1y$.

$$^-y\left(\frac{-2}{3}x\right) = {}^-1y\left(\frac{-2}{3}x\right)$$

$$= \left(^-1 \cdot \frac{-2}{3}\right)(yx)$$

$$= \frac{2}{3}yx \quad \text{or} \quad \frac{2}{3}xy$$

EXERCISE 4.16.1

Simplify:

1. $3\left(\dfrac{-7}{5}x\right)$

2. $\dfrac{-2}{9}\left(\dfrac{5}{4}y\right)$

3. $\left(\dfrac{7}{5}y\right) \cdot \dfrac{2}{7}$

4. $\dfrac{-1}{8}(^-8z)$

5. $5\left(\dfrac{1}{5}x\right)$

6. $\left(\dfrac{-3}{5}y\right) \cdot \dfrac{4}{7}$

7. $\left(\dfrac{-9}{10}z\right) \cdot \dfrac{-10}{9}$

8. $\dfrac{6}{7}\left(\dfrac{5}{6}w\right)$

9. $^-y\left(\dfrac{5}{9}x\right)$

10. $\left(\dfrac{-3}{4}y\right) \cdot {}^-x$

11. $7\left(\dfrac{-3}{4}\,x\right)$      16. $\dfrac{-3}{4}\left(\dfrac{5}{7}\,x\right)$

12. $\dfrac{-5}{3}\,(-6x)$      17. $\dfrac{-2}{5}\left(\dfrac{5}{3}\,x\right)$

13. $\left(\dfrac{4}{5}\,x\right)\cdot\dfrac{-7}{8}$      18. $\dfrac{-7}{10}\left(\dfrac{-10}{7}\,x\right)$

14. $\left(\dfrac{2}{7}\,x\right)\cdot\dfrac{7}{2}$      19. $\dfrac{x}{y}\left(\dfrac{-5y}{z}\right)$

15. $\dfrac{-1}{6}\,(-6x)$      20. $\dfrac{3x}{y}\left(\dfrac{-y}{3}\right)$

## 4.17   The Distributive Law of Multiplication over Addition

Besides the four laws of addition and the four laws of multiplication there is one law for the rational numbers which is concerned with both operations, addition and multiplication.

*The Distributive Law of Multiplication over Addition*

The product of a first rational number with the sum of two other rational numbers gives the same result as the sum of the products obtained by multiplying the first rational number by each of the other two. $x(y + z) = xy + xz$ becomes a true statement for any rational number replacements of the variables, $x$, $y$, and $z$.

The Distributive Law is used in two ways to simplify rational expressions. To remove parentheses from an expression the Distributive Law is used as follows:

(a) $$5\left(3x - \dfrac{1}{4}\right) = 5\left(3x + \dfrac{-1}{4}\right)$$

$$= 5\cdot 3x + 5\cdot\dfrac{-1}{4}$$

$$= 15x - \dfrac{5}{4}$$

(b)
$$\frac{-3}{7}\left(\frac{5}{3}x - \frac{7}{5}\right) = \frac{-3}{7}\left(\frac{5}{3}x + \frac{-7}{5}\right)$$

$$= \frac{-3}{7}\cdot\frac{5}{3}x + \frac{-3}{7}\cdot\frac{-7}{5}$$

$$= \frac{-5}{7}x + \frac{3}{5}$$

(c)
$$^-\left(4x - \frac{5}{2}\right) = {}^-1\left(4x - \frac{5}{2}\right)$$

$$= {}^-1\left(4x + \frac{-5}{2}\right)$$

$$= {}^-1\cdot 4x + {}^-1\cdot\frac{-5}{2}$$

$$= {}^-4x + \frac{5}{2}$$

To add rational expressions involving variables, the Distributive Law is used as follows:

(a)
$$5x - \frac{3}{4}x = 5x + \frac{-3}{4}x$$

$$= \left(5 + \frac{-3}{4}\right)x$$

$$= \frac{17}{4}x$$

(b)
$$\frac{-3}{5}x + \frac{2}{9}x = \left(\frac{-3}{5} + \frac{2}{9}\right)x$$

$$= \frac{-17}{45}x$$

(c)
$$\frac{8}{5}x - x = \frac{8}{5}x + {}^-1x$$

$$= \left(\frac{8}{5} + {}^-1\right)x$$

$$= \frac{3}{5}x$$

To simplify an expression such as $\dfrac{3}{4}\left(8x - \dfrac{5}{3}\right) - \dfrac{2}{7}\left(\dfrac{21}{4}x - 14\right)$,   the following steps are used:

$$\frac{3}{4}\left(8x - \frac{5}{3}\right) - \frac{2}{7}\left(\frac{21}{4}x - 14\right)$$

$$\frac{3}{4}\left(8x + \frac{{}^-5}{3}\right) + \frac{{}^-2}{7}\left(\frac{21}{4}x + {}^-14\right)$$

$$\frac{3}{4}\cdot 8x + \frac{3}{4}\cdot\frac{{}^-5}{3} + \frac{{}^-2}{7}\cdot\frac{21}{4}x + \frac{{}^-2}{7}\cdot {}^-14$$

$$6x + \frac{{}^-5}{4} + \frac{{}^-3}{2}x + 4$$

$$\left(6x + \frac{{}^-3}{2}x\right) + \left(\frac{{}^-5}{4} + 4\right)$$

$$\left(6 + \frac{{}^-3}{2}\right)x + \frac{11}{4}$$

$$\frac{9}{2}x + \frac{11}{4}$$

EXERCISE 4.17.1

1.   Remove the parentheses:

(a)  $6(x - 5)$

(c)  $\dfrac{{}^-2}{7}(3x - 5)$

(b)  $\dfrac{3}{4}(4x + 7)$

(d)  $\dfrac{4}{3}\left({}^-6x + \dfrac{5}{12}\right)$

(e) $-\left(\dfrac{-9}{5}x - 3\right)$

(h) $\dfrac{-3}{7}\left(7x - \dfrac{2}{3}\right)$

(f) $\dfrac{5}{8}\left(2x - \dfrac{4}{15}\right)$

(i) $\dfrac{-1}{9}(18x - 9)$

(g) $\dfrac{5}{6}\left(\dfrac{2}{3}x - \dfrac{1}{10}\right)$

(j) $\dfrac{1}{3}(5x - 6)$

2. Simplify:

(a) $5x + \dfrac{3}{4}x$

(g) $5\left(\dfrac{3}{5}x - \dfrac{2}{3}\right) - x$

(b) $\dfrac{-3}{4}x - x$

(h) $\dfrac{6}{5} - 3\left(\dfrac{2}{5}x - 1\right)$

(c) $\dfrac{5}{8}x - \dfrac{3}{2}x$

(i) $2\left(3x - \dfrac{5}{8}\right) - \dfrac{2}{5}(5x+10)$

(d) $\dfrac{4}{5}y + \dfrac{1}{3}y$

(j) $\dfrac{3}{2}(4x - 7) + 2\left(\dfrac{7}{2}x - \dfrac{11}{4}\right)$

(e) $\dfrac{-7}{2}x - \dfrac{7}{2}x$

(k) $3\left(\dfrac{5}{3}x - 2\right) + 7\left(3 - \dfrac{1}{5}x\right)$

(f) $\dfrac{2}{3}x - 6x$

(l) $\dfrac{4}{7}(-3x + 1) - (4 + x)$

## 4.18 Simplifying Complex Fractions

$\dfrac{\frac{5}{3}}{\frac{-8}{7}}$ and $\dfrac{\frac{-6}{5}}{\frac{1}{4}}$ are both complex fractions because the numerator

and denominator are both rational numbers.

The meaning of a complex fraction is closely related to the concept of

reciprocals. Recall that $\dfrac{5}{7}$ and $\dfrac{7}{5}$ are reciprocals because they have a product of 1. Another pair of reciprocals is $\dfrac{-8}{11}$ and $\dfrac{-11}{8}$ , because their product is 1.

In general, the reciprocal of $\dfrac{x}{y}$, where $x \neq 0$, and $y \neq 0$ is $\dfrac{y}{x}$ because $\dfrac{x}{y} \cdot \dfrac{y}{x} = 1$.

The rational number, $\dfrac{4}{9}$ , can be written as the multiplication expression, $4 \cdot \dfrac{1}{9}$ . 4 is the numerator of $\dfrac{4}{9}$ and $\dfrac{1}{9}$ is the reciprocal of the denominator, 9.

The rational number, $\dfrac{-7}{5}$ , can be written as the multiplication expression, $-7 \cdot \dfrac{1}{5}$ . $-7$ is the numerator of $\dfrac{-7}{5}$ and $\dfrac{1}{5}$ is the reciprocal of the denominator, 5.

Since $\dfrac{4}{9}$ is equal to $4 \cdot \dfrac{1}{9}$ and $\dfrac{-7}{5}$ is equal to $-7 \cdot \dfrac{1}{5}$ , it is possible to define $\dfrac{\frac{5}{3}}{\frac{-8}{7}}$ as $\dfrac{5}{3} \cdot \dfrac{-7}{8}$ . Notice that the complex fraction is written as a multiplication expression in which the numerator is multiplied by the reciprocal of the denominator.

The evaluation of the complex fraction, $\dfrac{\frac{-6}{5}}{\frac{1}{4}}$ , is accomplished in the following steps:

$$\dfrac{\frac{-6}{5}}{\frac{1}{4}} = \dfrac{-6}{5} \cdot \dfrac{4}{1} = \dfrac{-24}{5}$$

Similarly, $\dfrac{\dfrac{-5}{7}}{\dfrac{-15}{4}}$ , is evaluated as follows:

$$\frac{\dfrac{-5}{7}}{\dfrac{-15}{4}} = \frac{-5}{7} \cdot \frac{-4}{15} = \frac{4}{21}$$

EXERCISE 4.18.1

Evaluate:

1. $\dfrac{7}{\dfrac{3}{8}}$

2. $\dfrac{17}{\dfrac{5}{9}}$

3. $\dfrac{\dfrac{-3}{10}}{5}$

4. $\dfrac{\dfrac{2}{5}}{-3}$

5. $\dfrac{\dfrac{-7}{2}}{-6}$

6. $\dfrac{\dfrac{2}{3}}{\dfrac{5}{7}}$

7. $\dfrac{\dfrac{6}{5}}{\dfrac{2}{7}}$

8. $\dfrac{\dfrac{8}{3}}{\dfrac{-3}{5}}$

9. $\dfrac{\dfrac{-7}{8}}{\dfrac{-3}{4}}$

10. $\dfrac{\dfrac{-3}{7}}{\dfrac{11}{14}}$

11. $\dfrac{\dfrac{5}{3}}{\dfrac{3}{8}}$

16. $\dfrac{\dfrac{2x}{3}}{\dfrac{x}{y}}$

12. $\dfrac{10}{\dfrac{2}{5}}$

17. $\dfrac{\dfrac{-9x}{10}}{\dfrac{-9}{10}}$

13. $\dfrac{-7}{\dfrac{2}{9}}$

18. $\dfrac{\dfrac{-2x}{3y}}{\dfrac{1x}{6}}$

14. $\dfrac{\dfrac{-3}{4}}{\dfrac{5}{6}}$

19. $\dfrac{-5x}{\dfrac{5}{7}}$

15. $\dfrac{\dfrac{-7}{8}}{\dfrac{-7}{8}}$

20. $\dfrac{\dfrac{-7}{9}}{\dfrac{-2}{3}}$

## 4.19   Using Exponents with Rational Numbers

$-5x^2y^3z^4$ is an open expression using 2, 3, and 4 as exponents. $x^2$ means $xx$. $y^3$ means $yyy$. $z^4$ means $zzzz$.

Recall that $x^3 \cdot x^2 = xxx \cdot xx = x^5$. Similarly, $y^5 \cdot y = yyyyy \cdot y = y^6$.

Hence, $\dfrac{5x^2}{y} \cdot \dfrac{-3x^3}{y^5} = \dfrac{-15x^5}{y^6}$ .

$\left(\dfrac{-3x}{y}\right)^2$ means $\dfrac{-3x}{y} \cdot \dfrac{-3x}{y}$, and $\dfrac{-3x}{y} \cdot \dfrac{-3x}{y} = \dfrac{9x^2}{y^2}$ .

$\left(\dfrac{4x^2y}{z^3}\right)^3$ means $\dfrac{4x^2y}{z^3} \cdot \dfrac{4x^2y}{z^3} \cdot \dfrac{4x^2y}{z^3}$ , and this expression is simplified to

$\dfrac{64x^6y^3}{z^9}$ .

$\dfrac{10}{12}$ can be simplified to $\dfrac{5}{6}$ , because 2 is a factor of both 10 and 12.

$$\frac{10}{12} = \frac{5}{6} \cdot \frac{2}{2} = \frac{5}{6} \cdot 1 = \frac{5}{6}$$

$\dfrac{x^3}{x^5}$ can be simplified to $\dfrac{1}{x^2}$ , because $x^3$ is a factor of both $x^3$ and $x^5$.

$$\frac{x^3}{x^5} = \frac{1}{x^2} \cdot \frac{x^3}{x^3} = \frac{1}{x^2} \cdot 1 = \frac{1}{x^2}$$

$\dfrac{x^9}{x^4}$ can be simplified to $x^5$, because $x^4$ is a factor of both $x^9$ and $x^4$.

$$\frac{x^9}{x^4} = \frac{x^5}{1} \cdot \frac{x^4}{x^4} = \frac{x^5}{1} \cdot 1 = \frac{x^5}{1} = x^5$$

To simplify $\dfrac{-12x^5y^3}{8x^2y^7z}$ , the following steps are used:

$$\frac{-12x^5y^3}{8x^2y^7z} = \frac{-12}{8} \cdot \frac{x^5}{x^2} \cdot \frac{y^3}{y^7} \cdot \frac{1}{z}$$

$$= \frac{-3}{2} \cdot \frac{x^3}{1} \cdot \frac{1}{y^4} \cdot \frac{1}{z}$$

$$= \frac{-3x^3}{2y^4z}$$

EXERCISE 4.19.1

1.  Multiply:

(a)  $3x^2 \cdot {}^-7x$

(c)  $\dfrac{6x^5}{y} \cdot \dfrac{-2x^2}{3y^2}$

(b)  $(5xy^3z^4) \cdot (-x^3y^4z^2)$

(d)  $\dfrac{4y^3z}{9x^2} \cdot \dfrac{-3yz^5}{2x^3}$

(e) $\left(\dfrac{7x^2}{y^3}\right)^2$

(g) $\left(\dfrac{2x^4}{y}\right)^3$

(f) $\left(\dfrac{-9x^5}{4y^7}\right)^2$

(h) $\left(\dfrac{-3xy^2}{z^4}\right)^3$

2.  Simplify:

(a) $\dfrac{x^5}{x^2}$

(h) $\dfrac{-10x^9}{-2x^7}$

(b) $\dfrac{y^7}{y^{10}}$

(i) $\dfrac{14x^4}{21x}$

(c) $\dfrac{z^6}{z^3}$

(j) $\dfrac{10xy^7z^3}{4x^2y^3z}$

(d) $\dfrac{x^9}{x^2}$

(k) $\dfrac{-12x^5y^4z}{8x^2y^6z^4}$

(e) $\dfrac{-8z^5}{12z^3}$

(l) $\dfrac{-7x^5y^4z}{-14x^3z^3}$

(f) $\dfrac{-9x^9}{12x^4}$

(m) $\dfrac{-5x^3y^5z^6}{-7x^3y^2z^6}$

(g) $\dfrac{8x^4}{4x}$

(n) $\dfrac{-20xy^3z^5}{14x^3y^3z^3}$

## 4.20  Solving Equations of the Form $x + a = b$

To solve the equation, $x + \dfrac{3}{5} = \dfrac{1}{2}$, using the set of rational num-

bers as the domain means to find the set of all rational numbers that makes

$x + \dfrac{3}{5} = \dfrac{1}{2}$ a true statement. We say $\left\{\dfrac{-1}{10}\right\}$ is the truth set of $x + \dfrac{3}{5} = \dfrac{1}{2}$,

because $\dfrac{-1}{10} + \dfrac{3}{5} = \dfrac{1}{2}$ is a true statement.

Every rational number has an opposite, and the sum of any number and its opposite is zero. These facts lead to the easiest method for solving

equations such as $x + \dfrac{3}{5} = \dfrac{1}{2}$. Notice that $\dfrac{-3}{5}$ is the opposite of $\dfrac{3}{5}$.    When

$\dfrac{-3}{5}$ is added to both sides of the open sentence, $x + \dfrac{3}{5} = \dfrac{1}{2}$, the truth set

can easily be obtained by simplifying the expressions on both sides of the equation, as shown below:

$$x + \frac{3}{5} = \frac{1}{2}$$

$$\left(x + \frac{3}{5}\right) + \frac{-3}{5} = \frac{1}{2} + \frac{-3}{5}$$

$$x + \left(\frac{3}{5} + \frac{-3}{5}\right) = \frac{-1}{10}$$

$$x + 0 = \frac{-1}{10}$$

$$x = \frac{-1}{10}$$

Therefore, the truth set of $x + \dfrac{3}{5} = \dfrac{1}{2}$ is $\left\{\dfrac{-1}{10}\right\}$.

To solve $x - \dfrac{7}{8} = \dfrac{1}{4}$, first notice that $x - \dfrac{7}{8} = \dfrac{1}{4}$ means $x + \dfrac{-7}{8} = \dfrac{1}{4}$,

and the opposite of $\dfrac{-7}{8}$ is $\dfrac{7}{8}$. The solution of $x - \dfrac{7}{8} = \dfrac{1}{4}$ is shown below.

$$x - \frac{7}{8} = \frac{1}{4}$$

$$\left(x + \frac{-7}{8}\right) + \frac{7}{8} = \frac{1}{4} + \frac{7}{8}$$

$$x + \left(\frac{-7}{8} + \frac{7}{8}\right) = \frac{9}{8}$$

$$x + 0 = \frac{9}{8}$$

$$x = \frac{9}{8}$$

Therefore, the truth set of $x - \frac{7}{8} = \frac{1}{4}$ is $\left\{\frac{9}{8}\right\}$.

## EXERCISE 4.20.1

Find the truth sets.

1. $x + \frac{3}{4} = \frac{5}{6}$

2. $x - \frac{1}{3} = \frac{1}{2}$

3. $x + 3 = \frac{15}{4}$

4. $x - 2 = \frac{1}{5}$

5. $x + \frac{6}{7} = \frac{-1}{2}$

6. $x - \frac{9}{4} = {}^{-}1$

7. $z + \frac{10}{3} = 2$

8. $x - \frac{2}{5} = \frac{3}{7}$

9. $x + \frac{4}{3} = \frac{-9}{2}$

10. $x - \frac{9}{4} = \frac{7}{10}$

## 4.21   Solving Equations of the Form $ax = b$

Equations such as $x + \frac{2}{3} = \frac{1}{2}$ are solved by adding opposites.

$\dfrac{-2}{3}$ is the opposite of $\dfrac{2}{3}$. Since zero is the identity element for addition, the variable, $x$, is isolated on the left side of the equation as shown below:

$$\left(x + \frac{2}{3}\right) + \frac{-2}{3} = \frac{1}{2} + \frac{-2}{3}$$

$$x + 0 = \frac{-1}{6}$$

$$x = \frac{-1}{6}$$

In the equation, $\dfrac{3}{5}x = \dfrac{2}{7}$, the $\dfrac{3}{5}$ is multiplied by the variable $x$ in the open expression, $\dfrac{3}{5}x$. The identity element for multiplication is positive one, and the product of any rational number and its reciprocal is positive one. These facts are used in solving $\dfrac{3}{5}x = \dfrac{2}{7}$ by first multiplying both sides of the equation by the reciprocal of $\dfrac{3}{5}$, which is $\dfrac{5}{3}$. The solution of $\dfrac{3}{5}x = \dfrac{2}{7}$ is shown below:

$$\frac{3}{5}x = \frac{2}{7}$$

$$\frac{5}{3} \cdot \frac{3}{5}x = \frac{5}{3} \cdot \frac{2}{7}$$

$$1x = \frac{10}{21}$$

$$x = \frac{10}{21}$$

Therefore, the truth set of $\dfrac{3}{5}x = \dfrac{2}{7}$ is $\left\{\dfrac{10}{21}\right\}$.

To solve $^-5x = \dfrac{3}{8}$, the reciprocal of $^-5$, which is $\dfrac{^-1}{5}$, is used as follows:

$$^-5x = \frac{3}{8}$$

$$\frac{^-1}{5}\,(^-5x) = \frac{^-1}{5}\cdot\frac{3}{8}$$

$$\left(\frac{^-1}{5}\cdot{^-5}\right)x = \frac{^-3}{40}$$

$$1x = \frac{^-3}{40}$$

$$x = \frac{^-3}{40}$$

Therefore, the truth set of $^-5x = \dfrac{3}{8}$ is $\left\{\dfrac{^-3}{40}\right\}$.

EXERCISE 4.21.1

Find the truth sets:

1. $\dfrac{2}{9}x = \dfrac{3}{7}$

2. $\dfrac{^-5}{3}x = \dfrac{1}{2}$

3. $\dfrac{4}{5}x = \dfrac{3}{10}$

4. $\dfrac{^-6}{5}x = \dfrac{4}{15}$

5. $6x = \dfrac{^-4}{5}$

6. $^-3x = \dfrac{^-5}{7}$

7. $\dfrac{5}{9}x = \dfrac{2}{3}$

8. $\dfrac{^-4}{5}x = 7$

9. $11x = {}^-9$

10. $^-12x = 20$

### 4.22   Solving Equations of the Form $ax + b = c$

Equations such as $\dfrac{3}{5}x - \dfrac{1}{2} = \dfrac{5}{8}$ can be solved using the skills involved in the previous two sections. The first step in solving $\dfrac{3}{5}x - \dfrac{1}{2} = \dfrac{5}{8}$

is to add the opposite of $\dfrac{-1}{2}$ to both sides of the equation. After simplifying,

both sides are then multiplied by the reciprocal of $\dfrac{3}{5}$.   The   solution   of

$\dfrac{3}{5}x - \dfrac{1}{2} = \dfrac{5}{8}$ is accomplished in the following steps:

$$\frac{3}{5}x - \frac{1}{2} = \frac{5}{8}$$

$$\left(\frac{3}{5}x + \frac{-1}{2}\right) + \frac{1}{2} = \frac{5}{8} + \frac{1}{2}$$

$$\frac{3}{5}x = \frac{9}{8}$$

$$\frac{5}{3}\left(\frac{3}{5}x\right) = \frac{5}{3} \cdot \frac{9}{8}$$

$$x = \frac{15}{8}$$

Therefore, the truth set of $\dfrac{3}{5}x - \dfrac{1}{2} = \dfrac{5}{8}$ is $\left\{\dfrac{15}{8}\right\}$.

There are basically two steps involved in solving any equation of the form $ax + b = c$. The two steps are as follows:

1.   Add the opposite of $b$ to both sides of the equation, $ax + b = c$.

2.   After simplifying, multiply both sides of the equation by the reciprocal of the coefficient of the variable, $x$. The reciprocal

of $a$ is $\dfrac{1}{a}$.

As another example, the solution of $\dfrac{^-4}{7}x + \dfrac{3}{4} = \dfrac{1}{2}$ is shown below:

$$\frac{^-4}{7}x + \frac{3}{4} = \frac{1}{2}$$

$$\left(\frac{^-4}{7}x + \frac{3}{4}\right) + \frac{^-3}{4} = \frac{1}{2} + \frac{^-3}{4}$$

$$\frac{^-4}{7}x = \frac{^-1}{4}$$

$$\frac{^-7}{4}\left(\frac{^-4}{7}x\right) = \frac{^-7}{4} \cdot \frac{^-1}{4}$$

$$x = \frac{7}{16}$$

Therefore, the truth set of $\dfrac{^-4}{7}x + \dfrac{3}{4} = \dfrac{1}{2}$ is $\left\{\dfrac{7}{16}\right\}$ .

EXERCISE 4.22.1

Find the truth sets:

1.  $2x + 7 = 18$

2.  $3x - 6 = {}^-1$

3.  $6x + 5 = 4$

4.  $^-9x - 7 = 4$

5.  $^-3x + 2 = {}^-8$

6.  $11x - 6 = {}^-3$

7.  $^-3x + 4 = {}^-4$

8.  $15x - 4 = 9$

9.  $^-7x + 5 = {}^-12$

10.  $15 = 4x + 5$

11.  $\dfrac{3}{2}x + 4 = 9$

12.  $\dfrac{^-5}{4}x - 1 = 6$

13.  $\dfrac{1}{8}x + \dfrac{2}{3} = \dfrac{5}{6}$

14.  $\dfrac{3}{7}x - \dfrac{1}{4} = \dfrac{1}{12}$

15.  $\dfrac{2}{9}x + 1 = \dfrac{5}{3}$          18.  $\dfrac{-1}{4}x - \dfrac{3}{7} = \dfrac{5}{14}$

16.  $5x - \dfrac{1}{2} = 4$          19.  $\dfrac{7}{6}x + \dfrac{3}{10} = \dfrac{1}{5}$

17.  $\dfrac{2}{3}x + 2 = {}^-3$          20.  $\dfrac{-9}{5}x + \dfrac{1}{4} = \dfrac{1}{6}$

### 4.23  Solving More Involved Equations

The skills involved in solving equations such as $\dfrac{3}{4}x - \dfrac{5}{7} = \dfrac{4}{9}$  need only slight expansion to solve more involved equations such as $5(2x - 3) + 7x = 4(x + 2) - 3$. Only two new ideas are involved in solving such equations.

1.  Always simplify each side of the equation whenever such simplification is possible.

2.  Every open expression has an opposite. $4x$ and $^-4x$ are opposites, and $^-7x$ and $7x$ are opposites.

The solution of $5(2x - 3) + 7x = 4(x + 2) - 3$ is shown below:

$$5(2x - 3) + 7x = 4(x + 2) - 3$$

$$10x - 15 + 7x = 4x + 8 - 3$$

$$17x - 15 = 4x + 5$$

$$(17x - 15) + {}^-4x = (4x + 5) + {}^-4x$$

$$13x - 15 = 5$$

$$(13x - 15) + 15 = 5 + 15$$

$$13x = 20$$

$$\frac{1}{13}(13x) = \frac{1}{13} \cdot 20$$

$$x = \frac{20}{13}$$

Therefore, the truth set of $5(2x - 3) + 7x = 4(x + 2) - 3$ is $\left\{\dfrac{20}{13}\right\}$.

Notice that in the fourth line of the above example, $^-4x$ was added to both sides of the equation to eliminate the variable from the right side.

As another example, consider the equation,

$$x - 3(2x + 7) = 2(5x - 4) - 3x.$$

$$x - 6x - 21 = 10x - 8 - 3x$$

$$^-5x - 21 = 7x - 8$$

$$(^-5x - 21) - 7x = (7x - 8) - 7x$$

$$^-12x - 21 = ^-8$$

$$(^-12x - 21) + 21 = ^-8 + 21$$

$$^-12x = 13$$

$$\frac{^-1}{12}(^-12x) = \frac{^-1}{12} \cdot 13$$

$$x = \frac{^-13}{12}$$

Therefore, the truth set of $x - 3(2x + 7) = 2(5x - 4) - 3x$ is $\left\{ \dfrac{^-13}{12} \right\}$. Notice that in the fourth line of the above example, $^-7x$ is added to both sides of the equation to eliminate the variable from the right side.

EXERCISE 4.23.1

Find the truth sets:

1. $3(x - 2) + 5 = 7$
2. $4x - 3(x - 6) = 4$
3. $9x + 7 = 3x - 6$
4. $5x - 7 = 6 - x$
5. $x + 5 = 3x - 1$
6. $4x + 7 = x - 3$
7. $7 + 3(x - 2) = 2x - 3$
8. $2x + 3(x + 2) = 2(x - 5) + 8$
9. $4x - 3(x + 4) = 5 - 4x$
10. $3(2x - 4) + 7 = 4 - (x + 2)$
11. $3(5x - 4) - 6 = 2(x - 3)$
12. $(4x - 7) = 2(x - 3) + 5$

13.  $2(3x + 4) + 4 = 3(x - 5)$
14.  $9x - 5(2x - 3) = 8 + x$
15.  $4(x - 7) - 3(x - 6) = 5x + 1$
16.  $5(x - 8) - 2 = 19$
17.  $2(3x + 7) - 1 = 9$
18.  $5x - 3(x + 4) = 4x - 3$
19.  $3x + 8 = x - 7$
20.  $8x + 9 = 3x + 15$
21.  $5 - 3x = 8x + 7$
22.  $4 + 7x = 9 - 3x$
23.  $2(3x - 7) + 5 = x - 3$
24.  $4x - 3(2x - 1) = 5x + 1$
25.  $2(3x - 1) + 5 = 4 - (x + 5)$
26.  $9 - 3(x - 6) = 5x - (x + 6)$
27.  $7x + (3x - 5) = 2x - 1$
28.  $4(x - 1) + 3x = 1 - 5x$
29.  $2(x - 5) - 5(x + 4) = x - 1$
30.  $5(x + 1) - 3(2x + 7) = 4x + 3$

## 4.24  Solving Equations of the Form $\dfrac{a}{x} = b$

The equation $\dfrac{3}{x} = 7$ is solved by first multiplying both sides of the

equation by $x$ to eliminate the variable from the denominator.

$$\frac{3}{x} = 7$$

$$x \cdot \frac{3}{x} = x \cdot 7$$

$$3 = 7x$$

$$\frac{1}{7} \cdot 3 = \frac{1}{7}(7x)$$

$$\frac{3}{7} = x$$

Therefore, the truth set of $\dfrac{3}{x} = 7$, is $\left\{\dfrac{3}{7}\right\}$ .

The solution of $\dfrac{6}{x} = \dfrac{11}{x} + 7$ depends on the fact that $\dfrac{^{-}11}{x}$ is the oppo-

site of $\dfrac{11}{x}$ , and $\dfrac{6}{x} + \dfrac{^{-}11}{x}$ can be simplified to $\dfrac{^{-}5}{x}$ .

$$\frac{6}{x} = \frac{11}{x} + 7$$

$$\frac{6}{x} - \frac{11}{x} = \left(\frac{11}{x} + 7\right) - \frac{11}{x}$$

$$\frac{^{-}5}{x} = 7$$

$$x \cdot \frac{^{-}5}{x} = x \cdot 7$$

$$^{-}5 = 7x$$

$$\frac{1}{7} \cdot {^{-}5} = \frac{1}{7}\,(7x)$$

$$\frac{^{-}5}{7} = x$$

Therefore, the truth set of $\dfrac{6}{x} = \dfrac{11}{x} + 7$ is $\left\{\dfrac{^{-}5}{7}\right\}$ .

EXERCISE 4.24.1

Find the truth sets:

1. $\dfrac{2}{x} = {^{-}3}$ 

4. $\dfrac{5}{x} = {^{-}11}$

2. $\dfrac{^{-}3}{x} = 10$ 

5. $\dfrac{2}{x} - 3 = 4$

3. $\dfrac{^{-}4}{x} = {^{-}2}$ 

6. $7 - \dfrac{4}{x} = 6$

7. $\dfrac{4}{x} = \dfrac{6}{x} - 3$

14. $\dfrac{9}{x} = {}^{-}9$

8. $\dfrac{7}{x} = \dfrac{1}{x} + 4$

15. $\dfrac{3}{x} - 2 = 12$

9. $8 - \dfrac{3}{x} = 2 + \dfrac{5}{x}$

16. $8 - \dfrac{4}{x} = 5$

10. $\dfrac{2}{x} + 5 = 4 - \dfrac{3}{x}$

17. $\dfrac{2}{x} = \dfrac{9}{x} - 1$

11. $\dfrac{5}{x} = 7$

18. $\dfrac{3}{x} = 2 - \dfrac{6}{x}$

12. $\dfrac{2}{x} = {}^{-}8$

19. $\dfrac{3}{x} + 7 = \dfrac{9}{x} - 2$

13. $\dfrac{{}^{-}7}{x} = {}^{-}5$

20. $4 - \dfrac{5}{x} = \dfrac{4}{x} - 5$

## 4.25    Ordering the Rational Numbers

The rational numbers may be graphed on a number line like the one shown in Figure 4.1. Our concept of size related to the number line is

FIGURE 4.1.

that the numbers become greater as we go to the right. We say $\dfrac{5}{4}$ is greater

than 1, because $\dfrac{5}{4}$ is to the right of 1 in Figure 4.1. The symbol "$>$" is read

"greater than" and can be used to write the statement, $\dfrac{5}{4} > 1$. The symbol

"$<$" is read "less than" and can be used to write the true statement, $3 < 5$.

To graph $\dfrac{13}{5}$ on the number line it should be recognized that $\dfrac{13}{5}$ is equal

to $2 + \dfrac{3}{5}$. Therefore, $\dfrac{13}{5}$ is three-fifths of the way between 2 and 3, and is

graphed as shown in Figure 4.1.

To graph $\dfrac{^-11}{6}$ on the number line, notice that $\dfrac{^-11}{6}$ is $^-1 + \dfrac{^-5}{6}$. Hence,

the position of $\dfrac{^-11}{6}$ on the number line is between $^-1$ and $^-2$ as shown in

Figure 4.1.

EXERCISE 4.25.1

Graph each of the following rational numbers on a number line similar to figure 4.1.

1. $\dfrac{18}{7}$         6. $\dfrac{14}{3}$

2. $\dfrac{^-4}{3}$         7. $\dfrac{9}{4}$

3. $\dfrac{5}{2}$         8. $\dfrac{^-7}{2}$

4. $\dfrac{^-14}{5}$         9. $\dfrac{3}{10}$

5. $\dfrac{^-8}{9}$         10. $\dfrac{^-12}{7}$

## 4.26  Solving Inequalities

$3x - 5 > 7$ is an open sentence called an inequality. To simplify this inequality, the same procedure may be used as that employed with equations. Add opposites to isolate the variable term on one side of the inequality, and multiply by reciprocals to have the variable multiplied by 1.

The steps in simplifying the inequality, $3x - 5 > 7$, are as follows:

$$3x - 5 > 7$$

$$(3x - 5) + 5 > 7 + 5$$

$$3x > 12$$

$$x > 4$$

$3x - 5 > 7$ and $x > 4$ are equivalent inequalities because any number replacement that makes one of the statements true, also makes the other statement true.

There is one very important difference between solving equations and inequalities. Any non-zero number may be multiplied by both sides of an equation without destroying the equality, but this is not the case with inequalities. In fact, if a negative number is multiplied by both sides of an inequality the direction of the inequality is reversed. For example, $^-3 < {}^-1$, $3 > {}^-2$, and $1 < 4$ are all true statements. If $^-4$ is multiplied by both sides of each inequality the following statements are obtained:

| $^-3 < {}^-1$ | $3 > {}^-2$ | $1 < 4$ |
|---|---|---|
| $^-4 \cdot {}^-3 \ ? \ {}^-4 \cdot {}^-1$ | $^-4 \cdot 3 \ ? \ {}^-4 \cdot {}^-2$ | $^-4 \cdot 1 \ ? \ {}^-4 \cdot 4$ |
| $12 > 4$ | $^-12 < 8$ | $^-4 > {}^-16$ |

Notice that in each case the multiplication by a negative number resulted in a change of symbol from "$<$" to "$>$" or "$>$" to "$<$."

The following rule must be followed when multiplying both sides of an inequality by a non-zero number:

1. If $x > y$ and $z > 0$, then $xz > yz$. The direction of the inequality remains the same if the multiplier is positive.

2. If $x > y$ and $z < 0$, then $xz < yz$. The direction of the inequality is reversed if the multiplier is negative.

As an example, $^-3x < 5$ is solved as follows:

$$^-3x < 5$$

$$\frac{^-1}{3}(^-3x) > \frac{^-1}{3} \cdot 5$$

$$x > \frac{^-5}{3}$$

As another example, $4x + 7 > 6x - 2$ is solved as follows:

$$4x + 7 > 6x - 2$$

$$(4x + 7) - 6x > (6x - 2) - 6x$$

$$^{-}2x + 7 > ^{-}2$$

$$(^{-}2x + 7) - 7 > ^{-}2 - 7$$

$$^{-}2x > ^{-}9$$

$$\frac{^{-}1}{2}(^{-}2x) < \frac{^{-}1}{2} \cdot {^{-}9}$$

$$x < \frac{9}{2}$$

The solving of inequalities is accomplished in a very similar manner to the solving of equations with the following exception: When multiplying both sides of an inequality by a negative number, the result is a change of direction of the inequality.

EXERCISE 4.26.1

Find a simpler equivalent inequality for each of the following inequalities.

1.  $5x > 7$
2.  $^{-}3x > 11$
3.  $2x < ^{-}6$
4.  $^{-}4x < ^{-}9$
5.  $2x - 7 > 6$
6.  $3x + 8 < 6$
7.  $^{-}4x - 3 > 2$
8.  $^{-}7x + 9 < 3$
9.  $4x - 9 > x - 6$
10.  $2x + 3 < 5x - 6$
11.  $5x + 9 > 2$
12.  $3 - 6x < 17$
13.  $x - 3 < 4x + 1$
14.  $9 - x > x - 9$
15.  $5x - 3(2x - 1) > x - 6$
16.  $2(x - 1) - 3x < 5 - 7x$
17.  $4x - 7 < 2 + 3(x - 1)$
18.  $5(2x - 1) - 2(3x + 1) > x - 6$
19.  $2(3x + 1) - (4x + 5) > 6 - 5x$
20.  $3 - 2(5x - 6) < 6(x - 1) - 2x$

## 4.27   Properties of the System of Rational Numbers

The system of rational numbers is best summarized by listing its properties.

*Addition Properties:*

The Closure Law of Addition: The sum of any two rational numbers is always a rational number.

The Commutative Law of Addition: $x + y = y + x$

The Associative Law of Addition: $x + (y + z) = (x + y) + z$

Zero is the Identity Element for Addition: $x + 0 = x$

Every rational number has an Opposite: $x + {}^-x = 0$

*Multiplication Properties:*

The Closure Law of Multiplication: The product of two rational numbers is always a rational number.

The Commutative Law of Multiplication: $xy = yx$

The Associative Law of Multiplication: $x(yz) = (xy)z$

One is the Identity Element for Multiplication: $1x = x$

Every rational number except zero has a reciprocal: $x \cdot \dfrac{1}{x} = 1$

*A Property Involving both Addition and Multiplication:*

The Distributive Law of Multiplication over Addition:

$$x(y + z) = xy + xz$$

*Properties for Solving Open Sentences:*

(1) Any rational number may be added to both members of an equation to generate another equation with exactly the same truth set.

(2) Any non-zero rational number may be multiplied by both members of an equation to generate another equation with exactly the same truth set.

(3) Any rational number may be added to both members of an inequality to generate another inequality in the same direction with exactly the same truth set.

(4) Any positive rational number may be multiplied by both members of an inequality to generate another inequality in the same direction with exactly the same truth set.

(5) Any negative rational number may be multiplied by both members of an inequality to generate another inequality in the opposite direction with exactly the same truth set.

# 5 OPEN SENTENCES WITH TWO VARIABLES

## 5.1 Introduction

In previous chapters we have dealt with equations and inequalities involving a single variable. For example, $4x - 7 = 6x - 3$ and $3(x - 2) > 4 - x$ are two open sentences which involve a single variable, $x$. The first is an equation and the second is an inequality.

Whenever the variable in $4x - 7 = 6x - 3$ or $3(x - 2) > 4 - x$ is replaced by a rational number then a numerical statement is obtained. The statement may be true or it may be false depending upon which number is used as a replacement for the variable. For example, if $x$ is replaced by 5 in $4x - 7 = 6x - 3$ the numerical statement $4 \cdot 5 - 7 = 6 \cdot 5 - 3$ is obtained. This statement is false.

If $x$ is replaced by 5 in $3(x - 2) > 4 - x$ the numerical statement $3(5 - 2) > 4 - 5$ is obtained. This statement is true.

Any open sentence with one variable becomes a true or false numerical statement when that single variable is replaced by a rational number. A slightly different situation is present in the open sentences $4x + y = 7$ and $x - 2y > 3$. Each of these open sentences involves two variables. If just one variable is replaced by a rational number then the result is still an open sentence. For example, if $x$ is replaced by 3 in $4x + y = 7$ the open sentence $12 + y = 7$ is obtained. To obtain a numerical statement from

155

$4x + y = 7$ both variables must be replaced. Any rational number may replace $x$ and any rational number may replace $y$.

If $x$ is replaced by 5 and $y$ is replaced by $^-2$ in the equation $4x + y = 7$ the resulting statement is $4 \cdot 5 + ^-2 = 7$ or $18 = 7$ which is false. If $4x + y = 7$ is used with $x = 2$ and $y = ^-1$, the numerical statement $4 \cdot 2 + ^-1 = 7$ is obtained. This statement is true.

For the inequality $2x - y > 6$ the replacements $x = 4$ and $y = 3$ result in the numerical statement $2 \cdot 4 - 3 > 6$ or $5 > 6$ which is false. If $x = 1$ and $y = ^-5$ the inequality $2x - y > 6$ gives the numerical statement $2 \cdot 1 - ^-5 > 6$ or $7 > 6$ which is true.


EXERCISE 5.1.1

For each open sentence make the given replacements and determine the truth or falsity of the resulting numerical statement.

1.  $2x + y = 9$
    (a) $x = 6, y = ^-2$      (c) $x = 7, y = ^-5$
    (b) $x = 4, y = 1$      (d) $x = ^-2, y = 13$

2.  $3x - y = 4$
    (a) $x = 2, y = 2$      (c) $x = ^-1, y = ^-7$
    (b) $x = 5, y = 4$      (d) $x = 4, y = 6$

3.  $3x - 4y = 7$
    (a) $x = 5, y = 1$      (c) $x = 5, y = 2$
    (b) $x = 4, y = ^-1$      (d) $x = 1, y = ^-1$

4.  $5x + 2y = 4$
    (a) $x = 0, y = 2$      (c) $x = \dfrac{4}{5}, y = 0$

    (b) $x = 1, y = \dfrac{^-1}{2}$      (d) $x = ^-1, y = 6$

5.  $2x + y > 7$
    (a) $x = 5, y = 1$      (c) $x = 4, y = ^-3$
    (b) $x = ^-2, y = 9$      (d) $x = 3, y = 5$

6.  $3x - 2y < 5$
    (a) $x = 4, y = 3$      (c) $x = 3, y = ^-4$
    (b) $x = ^-5, y = ^-6$      (d) $x = ^-6, y = 2$

7.  $4x - y > 8$
    (a) $x = 2, y = 3$      (c) $x = ^-4, y = ^-10$
    (b) $x = 8, y = 10$      (d) $x = 0, y = ^-10$

8.  $x - 5y < 4$
    (a) $x = 6, y = 1$      (c) $x = 9, y = ^-1$
    (b) $x = ^-3, y = ^-1$      (d) $x = 15, y = 2$

## 5.2  Ordered Pairs as Replacements

An open sentence in two variables requires a pair of rational number replacements before a numerical statement can be obtained. In the last exercise replacements for both variables were given by listing them $x = 5$ and $y = 2$. There is a simpler method for designating the number replacements to be used in an open sentence involving two variables. $(6, {}^-3)$ is an ordered pair involving the two numbers 6 and $^-3$. If we agree to use ordered pairs $(x, y)$ so that the first number listed shows the replacement for $x$ and the second number shows the replacement for $y$, then the ordered pair $(6, {}^-3)$ and the equality $x + y = 7$ results in the numerical statement $6 + {}^-3 = 7$, which is false.

Using the open sentence $3y - 2x < 7$ the ordered pair $(4, 5)$ results in the statement $3 \cdot 5 - 2 \cdot 4 < 7$. Notice that the ordered pair $(4, 5)$ designates 4 as the replacement for $x$ and 5 as the replacement for $y$ regardless of the position of these variables in the open sentence.

EXERCISE 5.2.1

Use the ordered pairs $(x, y)$ to write a numerical statement from each open sentence and determine the truth or falsity of each such statement.

1.   $2x + y = 7$,        $(5, {}^-3)$
2.   $x - 4y = 6$,        $(2, 1)$
3.   $5x + 2y = {}^-6$,        $({}^-4, 7)$
4.   $3x - y = 2$,        $({}^-1, {}^-5)$
5.   $4x + 2y > 5$,        $(3, {}^-3)$
6.   $x - y < 6$,        $(4, {}^-2)$
7.   $x - 7y > 24$,        $({}^-3, {}^-4)$
8.   $3x - 2y < 9$,        $(9, 7)$
9.   $4 - 2y = 3x$,        $({}^-2, 5)$
10.   $1 + 3y = 5x$        $({}^-1, {}^-2)$
11.   $2y - 3 = 4x$        $(5, 4)$
12.   $x + 5 > 2y$,        $({}^-3, 1)$
13.   $4y - 3x < 7$,        $(7, 1)$
14.   $7 - 2y < 5x$,        $(2, 6)$
15.   $5y + 2x - 1 > 0$        $(4, 1)$
16.   $4x - 7y = 25$        $(8, 1)$
17.   $x - y = 10$        $(7, {}^-3)$
18.   $y = 5x - 3$        $(0, 2)$
19.   $3x + y = 5x - 2$        $({}^-3, {}^-8)$
20.   $y - 4x = 2y + 6$        $({}^-5, 12)$

### 5.3  Solutions for Open Sentences in Two Variables

(3, 1) is a *solution* for the equation $2x + y = 7$ because $2 \cdot 3 + 1 = 7$ is a true statement.

(5, ⁻4) is not a *solution* for the equation $2x + y = 7$ because $2 \cdot 5 + ⁻4 = 7$ is a false statement.

(6, ⁻2) is a *solution* for $x - y > 5$ because $6 - ⁻2 > 5$ is a true statement.

(10, 6) is not a *solution* for $x - y > 5$ because $10 - 6 > 5$ is a false statement.

An ordered pair $(a, b)$ is a solution for an open sentence in two variables whenever the statement obtained by replacing the variables is true. An ordered pair $(a, b)$ is not a solution for an open sentence in two variables whenever the statement obtained by replacing the variables is false.

EXERCISE 5.3.1

For each open sentence determine those ordered pairs that are solutions.

1.  $5x - y = 6$,      (2, 4), (⁻1, ⁻11), (3, 7), (0, ⁻6)

2.  $2x + 3y = 7$,      $(1, 4), (5, ⁻1), \left(0, \dfrac{7}{3}\right), (⁻1, 3)$

3.  $4x - 3y = ⁻6$,      (0, 2), (1, 3), (2, 5), (3, 6)

4.  $x + 3y = 1$,      $(⁻2, 1), (7, ⁻2), (⁻5, 2), \left(0, \dfrac{1}{3}\right)$

5.  $2x + y = 5$,      (⁻6, 17), (0, 5), (3, 1), (5, ⁻5)
6.  $x + 5y > 4$,      (6, 1), (10, ⁻1), (12, ⁻2), (0, 2)
7.  $2x - y < 5$,      (6, 1), (4, 5), (8, 8), (2, 2)
8.  $2x - 3y > 0$,      (⁻2, ⁻4), (5, 2), (⁻4, 1), (6, 4)
9.  $x + 4y < 0$,      (5, ⁻2), (⁻4, ⁻1), (⁻2, ⁻1), (4, 1)
10.  $3x + 2y > 5$,      (⁻5, 8), (5, ⁻3), (3, ⁻5), (1, 1)

### 5.4  Finding Solutions for Open Sentences in Two Variables

There is no limit to the number of ordered pair solutions for $3x - y = 7$. (1, ⁻4), (2, ⁻1), and (3, 2) are three such solutions, but in this section we are interested in finding any number of solutions desired.

Any rational number may replace $x$ in $3x - y = 7$ and this fact is used to generate solutions for the equation. If $x$ is replaced by 2 in $3x - y = 7$ the open sentence $6 - y = 7$ is obtained. This equation can be solved to show $y$ must be replaced by ⁻1. Hence, (2, ⁻1) is a solution for $3x - y = 7$.

To find another solution for $3x - y = 7$ select any rational number as a replacement for $x$ and solve the resultant equation to find the associated value for $y$.

$$3x - y = 7 \qquad \text{Let } x = 5$$
$$3 \cdot 5 - y = 7$$
$$^-y = ^-8$$
$$y = 8$$

Hence $(5, 8)$ is a solution for $3x - y = 7$.

An ordered pair solution for $2x - 3y = 8$ can be found by choosing any rational number as the replacement for $x$ and solving the resultant equation.

$$2x - 3y = 8 \qquad \text{Let } x = 2$$
$$2 \cdot 2 - 3y = 8$$
$$4 - 3y = 8$$
$$^-3y = 4$$
$$y = \frac{^-4}{3}$$

Hence $\left(2, \dfrac{^-4}{3}\right)$ is a solution for $2x - 3y = 8$.

To complete the ordered pair $(3, \underline{\phantom{xx}})$ as a solution for $x + 5y = 9$ the following steps are used.

$$x + 5y = 9 \qquad (3, \underline{\phantom{xx}})$$
$$3 + 5y = 9$$
$$5y = 6$$
$$y = \frac{6}{5}$$

Hence, $\left(3, \dfrac{6}{5}\right)$ is a solution for $x + 5y = 9$.

EXERCISE 5.4.1

Complete the ordered pair to obtain a solution for the equation.

1. $(2, \underline{\phantom{xx}})$   for   $2x + 3y = 11$
2. $(^-1, \underline{\phantom{xx}})$   for   $x - 6y = 9$

3. $(6, \underline{\quad})$  for  $x + 2y = {}^-3$
4. $({}^-3, \underline{\quad})$  for  $2x - 5y = 4$
5. $(4, \underline{\quad})$  for  $3x + 2y = 8$
6. $(5, \underline{\quad})$  for  $2x + 7y = 17$
7. $({}^-2, \underline{\quad})$  for  $4x - 3y = 11$
8. $(7, \underline{\quad})$  for  $x + 4y = 1$
9. $(\underline{\quad}, 2)$  for  $x - 3y = 5$
10. $(\underline{\quad}, {}^-3)$  for  $2x + 5y = {}^-3$
11. $(3, \underline{\quad})$  for  $7x - y = 19$
12. $({}^-1, \underline{\quad})$  for  $3x + 2y = 7$
13. $(0, \underline{\quad})$  for  $5x - 3y = 2$
14. $(2, \underline{\quad})$  for  $x + 5y = 2$
15. $({}^-3, \underline{\quad})$  for  $2x - y = 1$
16. $(4, \underline{\quad})$  for  $2x - 2y = 5$
17. $(7, \underline{\quad})$  for  $3x - 4y = 9$
18. $({}^-2, \underline{\quad})$  for  $5x - y = 6$
19. $(\underline{\quad}, 7)$  for  $2x + y = 11$
20. $(\underline{\quad}, {}^-3)$  for  $x - y = 6$

## 5.5  Solving Equations of the Form $ax + by = c$ **for** $y$

Quite often we desire to find two or more solutions for an equation with two variables. It is fairly easy to find solutions by the method of the last exercise, but by changing the form of an equation solutions are sometimes easier to find.

To solve $4x - 3y = 7$ for $y$ the concepts of opposites and reciprocals are of primary importance. The first step in solving $4x - 3y = 7$ for $y$ is to add $^-4x$ to both sides of the equation. After simplifying, both sides are

multiplied by $\dfrac{^-1}{3}$ to isolate $y$ on the left side of the equality.

$$4x - 3y = 7$$

$$^-4x + (4x - 3y) = {}^-4x + 7$$

$$^-3y = {}^-4x + 7$$

$$\frac{^-1}{3} \cdot {}^-3y = \frac{^-1}{3} ({}^-4x + 7)$$

$$y = \frac{4}{3}x - \frac{7}{3}$$

To solve $2x + 5y = 10$ for $y$ the opposite of $2x$ is first added to both sides. After simplifying, $\dfrac{1}{5}$ is multiplied by both members of the equation to isolate $y$ on the left side of the equality.

$$2x + 5y = 10$$

$$^{-}2x + (2x + 5y) = ^{-}2x + 10$$

$$5y = ^{-}2x + 10$$

$$\frac{1}{5} \cdot 5y = \frac{1}{5}\,(^{-}2x + 10)$$

$$y = \frac{^{-}2}{5}\,x + 2$$

EXERCISE 5.5.1

Solve each equation for $y$.

| | |
|---|---|
| 1.  $x + y = 7$ | 11.  $3x - 5y = 8$ |
| 2.  $x + 2y = ^{-}1$ | 12.  $6x - 2y = ^{-}4$ |
| 3.  $x - y = 11$ | 13.  $4x + 2y = 7$ |
| 4.  $3x + y = 17$ | 14.  $3x + 7y = ^{-}2$ |
| 5.  $2x - y = 4$ | 15.  $3x - 5y = 4$ |
| 6.  $x + 5y = 7$ | 16.  $x - 2y = 7$ |
| 7.  $4x + y = 9$ | 17.  $3x + 5y = 10$ |
| 8.  $x + 4y = 15$ | 18.  $2x - 7y = 6$ |
| 9.  $x - 3y = 11$ | 19.  $4x - 3y = 9$ |
| 10.  $3x - 7y = 14$ | 20.  $2x + 4y = 8$ |

## 5.6   Finding Solutions for Equations of the Form $y = ax + b$

Given an equation such as $y = 3x + 2$ a number of solutions can be easily found because whenever $x$ is replaced by a rational number the right member becomes a numerical expression equal to $y$.

For example, if $x$ is replaced by 5 in $y = 3x + 2$,

$$y = 3x + 2$$

$$y = 3 \cdot 5 + 2$$

$$y = 17$$

Hence $(5, 17)$ is a solution for $y = 3x + 2$.

If $x$ is replaced by zero in $y = 3x + 2$ the solution is easily obtained as

$$y = 3x + 2$$
$$y = 3 \cdot 0 + 2$$
$$y = 2$$

Hence $(0, 2)$ is another solution for $y = 3x + 2$.

The ease with which solutions for $y = \dfrac{-3}{5} x + 4$ are found is directly traceable to the fact that every replacement for $x$ requires only the evaluation of a numerical expression to give the corresponding value for $y$.

If $x = 5$ then $y = \dfrac{-3}{5} x + 4$ becomes

$$y = \frac{-3}{5} \cdot 5 + 4$$
$$y = {}^-3 + 4$$
$$y = 1$$

Hence $(5, 1)$ is a solution for $y = \dfrac{-3}{5} x + 4$.

If $x = 10$ then $y = \dfrac{-3}{5} x + 4$ becomes

$$y = \frac{-3}{5} \cdot 10 + 4$$
$$y = {}^-6 + 4$$
$$y = {}^-2$$

Hence $(10, {}^-2)$ is a solution for $y = \dfrac{-3}{5} x + 4$.

Notice that for the equation $y = \dfrac{-3}{5} x + 4$ any replacement for $x$ that is a number evenly divisible by 5 results in a relatively simple numerical expression because the coefficient of $x$ has the denominator 5.

For the equation $y = \dfrac{2}{7} x - 3$ any number divisible by 7 would be an easy replacement for $x$ because it will simplify the arithmetic necessary to find solutions for the equation.

$$y = \frac{2}{7}x - 3 \qquad\qquad \text{Let} \quad x = 7$$

$$y = \frac{2}{7} \cdot 7 - 3 = {}^-1 \qquad\qquad (7, {}^-1)$$

$$\text{Let} \quad x = 14$$

$$y = \frac{2}{7} \cdot 14 - 3 = 1 \qquad\qquad (14, 1)$$

$$\text{Let} \quad x = 0$$

$$y = \frac{2}{7} \cdot 0 - 3 = {}^-3 \qquad\qquad (0, {}^-3)$$

$(7, {}^-1)$, $(14, 1)$, and $(0, {}^-3)$ are three solutions of $y = \frac{2}{7}x - 3$ that are easily found using the equation in this form.

EXERCISE 5.6.1

1.  Find any three solutions for each equation.

    (a) $y = 5x - 6$           (e) $y = \frac{5}{2}x - 4$

    (b) $y = {}^-2x + 7$           (f) $y = \frac{{}^-7}{3}x + 2$

    (c) $y = \frac{2}{3}x + 1$           (g) $y = \frac{7}{4}x - 6$

    (d) $y = \frac{{}^-3}{4}x - 6$           (h) $y = \frac{{}^-1}{6}x + 5$

2.  Solve each equation for $y$ and then find any three solutions for it.

    (a) $2x + y = 7$           (e) $3x + 2y = 8$

    (b) $3x - y = 6$           (f) $4x - 3y = 3$

    (c) $4x + 2y = 10$         (g) $2x + 7y = {}^-14$

    (d) $x - y = 7$           (h) $5x - 6y = 18$

### 5.7  Comparing Rates of Change

For the equation $y = 3x$ if $x$ is replaced by 1 then $(1, 3)$ is the solution and if $x$ is replaced by 2 then $(2, 6)$ is the solution.

The two solutions $(1, 3)$ and $(2, 6)$ for the equation $y = 3x$ show that when the replacement for $x$ is increased by 1 then the matching replacement for $y$ must be increased by 3.

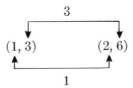

This increase in $y$ for any increase in $x$ can readily be observed from the equation $y = 3x$. Notice that the coefficient of $x$ is 3. Hence whenever $x$ is increased by 1, $y$ must show an increase of 3. Similarly, when $x$ is increased by 5 then $y$ must show an increase of $3 \cdot 5$ or 15.

Hence $(1, 3)$ and $(6, 18)$ are both solutions of $y = 3x$.

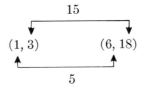

The preceding discussion of how the value of $y$ increases as the value of $x$ increases for the equation $y = 3x$ works equally well for the equation $y = 2x + 7$. At first glance it may seem that the 7 in $y = 2x + 7$ will have a direct effect on the manner in which $y$ increases for any increase in $x$, but this is not the case.

$(1, 9)$ and $(2, 11)$ are both solutions for $y = 2x + 7$. Notice that as the value of $x$ increases by 1 then the value of $y$ increases by 2.

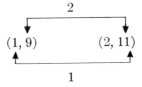

Similarly, as $x$ is increased by 5 then $y$ must increase by $2 \cdot 5$ or 10. This is because the coefficient of $x$ is 2 in $y = 2x + 7$. Both $(1, 9)$ and $(6, 19)$ are solutions of $y = 2x + 7$.

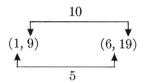

For the equation $y = {}^-4x + 5$ the coefficient of $x$ is a negative number, $^-4$. Therefore as the value of $x$ is increased by 1 then the value of $y$ must be decreased by 4. Both $(1, 1)$ and $(2, {}^-3)$ are solutions for $y = {}^-4x + 5$.

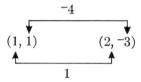

If $x$ is increased by 5 then $y$ must decrease by 20 for solutions of $y = {}^-4x + 5$ because $^-4$ is the coefficient of $x$. Both $(1, 1)$ and $(6, {}^-19)$ are solutions of $y = {}^-4x + 5$.

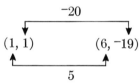

EXERCISE 5.7.1

1.  For each open sentence state the change in $y$ whenever $x$ is increased by 1.

    (a) $y = 4x$

    (b) $y = 5x - 2$
    (c) $y = {}^-3x + 7$

    (d) $y = 2x - 5$

    (e) $y = \dfrac{4}{5}x + 3$

    (f) $y = \dfrac{{}^-2}{3}x - 1$

    (g) $y = 3x + 7$
    (h) $y = {}^-x - 1$

    (i) $y = \dfrac{2}{5}x - 2$

    (j) $y = \dfrac{{}^-7}{4}x + 5$

2.  For each open sentence state the change in $y$ for the given increase in $x$.
    (a) $y = 7x - 2$   when $x$ increases by 4
    (b) $y = {}^-6x + 1$   when $x$ increases by 2
    (c) $y = {}^-3x + 8$   when $x$ increases by 5
    (d) $y = 5x - 7$   when $x$ increases by 3

(e)  $y = \dfrac{3}{4}x - 1$   when $x$ increases by 4

(f)  $y = \dfrac{-2}{5}x + 4$   when $x$ increases by 5

(g)  $y = \dfrac{5}{3}x - 1$   when $x$ increases by 3

(h)  $y = \dfrac{-5}{9}x + 7$   when $x$ increases by 9

(i)  $y = \dfrac{7}{2}x - 3$   when $x$ increases by 2

(j)  $y = \dfrac{-1}{3}x + 5$   when $x$ increases by 3

### 5.8  Graphing Equations with Two Variables

Figure 5.1 below shows a pair of perpendicular lines that may be used to graph ordered pairs of rational numbers and pictorially represent equations such as $2x - y = 7$.

The two lines have been labeled as the $x$ axis and $y$ axis in keeping with our use of ordered pairs of the form $(x, y)$. The location of seven ordered pairs is shown in figure 5.1. Notice that in each case the first number

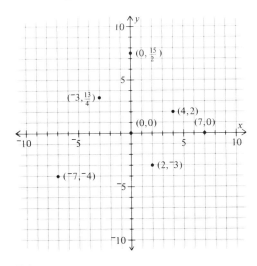

FIGURE 5.1

in the ordered pair designates the horizontal distance and direction from the origin, (0, 0). The second number of the ordered pair designates the vertical distance and direction from the origin, (0, 0).

EXERCISE 5.8.1

1. State the ordered pair of rational numbers that should name each point labeled by a capital letter in figure 5.2.

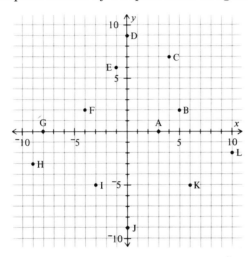

FIGURE 5.2

2. Graph the following ordered pairs on figure 5.3.
   (a) (6, 5)
   (b) (⁻3, 0)
   (c) (⁻4, 7)
   (d) (5, 0)
   (e) (4, ⁻6)
   (f) (⁻3, ⁻3)
   (g) (0, ⁻7)
   (h) (⁻8, 1)
   (i) (⁻2, 0)
   (j) (⁻6, ⁻9)

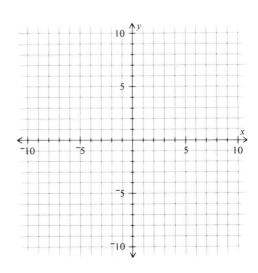

FIGURE 5.3

### 5.9 Straight Lines as Graphs of Equations of the Form $ax + by = c$

Any equation in two variables of the form $ax + by = c$ may be pictorially represented by a straight line on a graph. The straight line representing the solutions of $2x - 3y = 6$ is shown in figure 5.4.

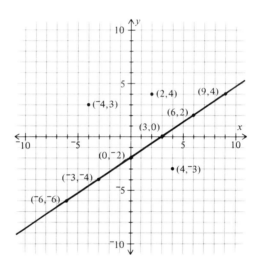

FIGURE 5.4

Figure 5.4 shows the straight line representing the solutions of $2x - 3y = 6$. Six specific points have been marked on the straight line. Notice that each of these six points has an ordered pair that is a solution for $2x - 3y = 6$. Three other points not on the line of $2x - 3y = 6$ have also been marked. They are $(^-4, 3)$, $(2, 4)$, and $(4, ^-3)$. Notice that these three ordered pairs are not solutions of $2x - 3y = 6$.

Every equation of the form $ax + by = c$ where $b \neq 0$ or $a \neq 0$ has a graph which is a straight line. Those points on the line represent ordered pairs which are solutions for the equation. Those points not on the line represent ordered pairs which are not solutions for the equation.

EXERCISE 5.9.1

Each problem shows the graph of the equation of a straight line. Locate each of the given ordered pairs on the graph and determine which ones are solutions of the equation. Check your answers using the equation.

1.  (5, 2)
    (1, 4)
    (−2, −1)
    (−4, −1)
    (5, −2)

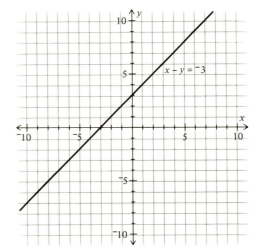

2.  (4, −3)
    (6, 4)
    (−2, 6)
    (3, 0)
    (0, 3)

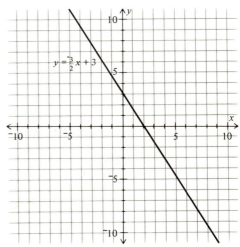

3.  (6, −2)
    (−2, −3)
    (4, 0)
    (0, 2)
    (−4, 6)

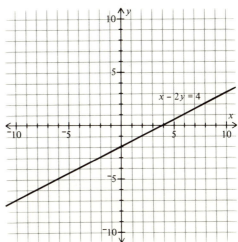

4.  $(4, ^-4)$
    $(1, ^-4)$
    $(4, 2)$
    $(0, ^-6)$
    $(^-3, 0)$

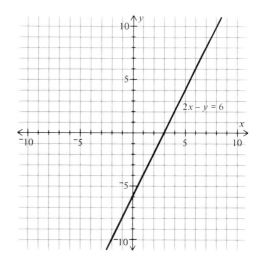

5.  $(3, ^-2)$
    $(^-6, 1)$
    $(0, 3)$
    $(6, ^-3)$
    $(^-5, ^-2)$

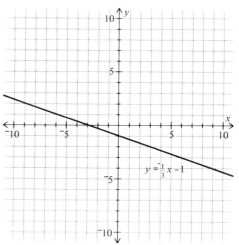

## 5.10   Graphing Solutions for Linear Equations

We have now seen that every equation of the form $ax + by = c$ has a graph which is a straight line. Every point on the line represents an ordered pair solution of the equation and every point not on the line does not represent an ordered pair solution of the equation.

To graph the equation $x + y = 2$ it is necessary to understand that we are looking for a straight line. Since two points determine a unique straight line and any solution of $x + y = 2$ represents a point on the line of the graph, it is only necessary to find two solutions of $x + y = 2$.

Choose any two rational numbers to replace $x$ in $x + y = 2$. Suppose 0 and 5 are chosen as replacements for $x$.

If $x = 0$ then $x + y = 2$

$$0 + y = 2$$

$$y = 2$$

(0, 2) is one solution.

If $x = 5$ then $x + y = 2$

$$5 + y = 2$$

$$y = {}^-3$$

(5, $^-$3) is another solution.

Graph the two solutions, (0, 2) and (5, $^-$3), as shown in figure 5.5.

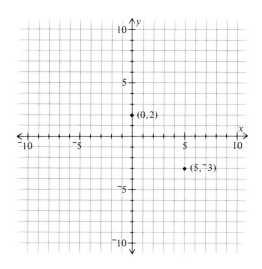

FIGURE 5.5

Once two solutions of an equation have been graphed then a straight line can be drawn through them. The two points graphed in figure 5.5 lead to the completed graph of $x + y = 2$ shown in figure 5.6.

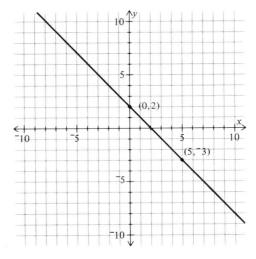

FIGURE 5.6

To check the graph of figure 5.6 choose any third point on the line of $x+y=2$ and make certain its ordered pair is a solution of the equation.

To find the straight line graph of any equation of the form $ax + by = c$:

(1)    Find two ordered pair solutions for the equation.

(2)    Graph the two points obtained in step 1.

(3)    Draw the straight line through the two points graphed in step 2.

EXERCISE 5.10.1

Find the straight line graph for each equation.

1.  $x + y = {}^-3$                9.   $2x + 4y = 7$
2.  $2x + y = 6$               10.  $5x - 2y = 4$
3.  $x - 2y = 4$               11.  $y = {}^-2x + 4$
4.  $x + 3y = 9$               12.  $y = 3x - 5$

5.  $3x + y = {}^-2$            13.  $y = \dfrac{1}{2}x + 1$

6.  $4x - y = 5$               14.  $y = \dfrac{-3}{5}x - 4$

7.  $2x + 3y = 8$              15.  $y = \dfrac{5}{4}x + 3$
8.  $3x - 4y = 5$

## 5.11   Graphing Equations Using the Form $y = mx + b$

In the last exercise equations were graphed by finding two solutions and drawing the straight line through the two points. When an equation is of the form $y = mx + b$ such as $y = 3x + 4$ and $y = {}^{-}2x + 1$ there is another, possibly easier, method for graphing the straight line.

$y = 3x + 4$ is an equation of the form $y = mx + b$ where $m = 3$ and $b = 4$. For $y = 3x + 4$ it is easy to find the point where the line crosses the $y$ axis because anywhere along the $y$ axis the value of $x$ must be zero. If $x$ is replaced by zero in $y = 3x + 4$ then $y = 3 \cdot 0 + 4$ or $y = 4$ is obtained. We say 4 is the $y$ *intercept* of $y = 3x + 4$ because the graph must cross the $y$ axis at $(0, 4)$.

Whenever an equation is of the form $y = mx + b$ then $b$ is the $y$ *intercept*. If $x$ is replaced by zero in $y = mx + b$ then $y = m \cdot 0 + b$ or $y = b$ is obtained. Hence $(0, b)$ is a solution of $y = mx + b$ and the graph crosses the $y$ axis at $(0, b)$.

For the equation $y = 5x - 6$, $^{-}6$ is the $y$ intercept because if $x$ is replaced by zero then $y = 5 \cdot 0 - 6$ or $y = {}^{-}6$ is obtained. Hence the graph of $y = 5x - 6$ crosses the $y$ axis at $(0, {}^{-}6)$.

For the equation $y = 2x + \dfrac{3}{4}$, $\dfrac{3}{4}$ is the $y$ intercept. In this case if $x$ is replaced by zero, then $y = 2 \cdot 0 + \dfrac{3}{4}$ or $y = \dfrac{3}{4}$ is the result. The graph of $y = 2x + \dfrac{3}{4}$ crosses the $y$ axis at $\left(0, \dfrac{3}{4}\right)$.

EXERCISE 5.11.1

Find the $y$ intercept for the graph of each equation.

1.  $y = 2x + 3$      4.  $y = \dfrac{2}{3}x - 4$

2.  $y = {}^{-}5x - 5$      5.  $y = 13x$

3.  $y = {}^{-}x + 1$      6.  $y = 4x - \dfrac{5}{6}$

7.  $y = \dfrac{-7}{4} x + \dfrac{2}{3}$ 

9.  $y = 3x + 6$

8.  $y = \dfrac{3}{4} x - \dfrac{2}{7}$ 

10.  $y = \dfrac{-2}{3} x + \dfrac{7}{5}$

EXERCISE 5.11.2

Solve each equation for $y$. Then find its $y$ intercept.

1.  $x + y = 7$ 
2.  $3x - 2y = 8$ 
3.  $4x + 3y = {}^{-}1$ 
4.  $x - 4y = 0$ 
5.  $x + 3y - 4 = 0$ 

6.  $2x - 7y = 6$ 
7.  $3x - 3y = 4$ 
8.  $x + 6y = 0$ 
9.  $x - 4y = 6$ 
10.  $3x + 2y = {}^{-}9$

One solution of $y = 3x - 5$ is $(0, {}^{-}5)$ because ${}^{-}5$ is the $y$ intercept. Another point on the graph may be found using the fact that as the values of $x$ increase then the values of $y$ must also increase three times as quickly. Recall that in the equation $y = 3x - 5$ the coefficient of $x$ is 3 and therefore as the value for $x$ is increased by 1 then the corresponding value for $y$ is increased by $3 \cdot 1$ or 3.

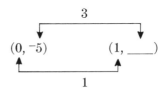

FIGURE 5.7

Hence the blank in figure 5.7 must be replaced by ${}^{-}2$. This provides us with two solutions, $(0, {}^{-}5)$ and $(1, {}^{-}2)$, for $y = 3x - 5$ and therefore the graph of the line can be drawn through the two points. Notice that these two points were obtained using only the numbers in the equation $y = 3x - 5$.

The ${}^{-}5$ in $y = 3x - 5$ is the $y$ intercept and gives us the point $(0, {}^{-}5)$.

The 3 in $y = 3x - 5$ is called the *slope* and because it forces $y$ to increase 3 whenever $x$ is increased by 1 we can obtain the second point $(1, {}^{-}2)$.

To graph $y = \dfrac{-3}{4}x + 2$ the $y$ intercept and slope are used as follows:

(1)  The $y$ intercept of $y = \dfrac{-3}{4}x + 2$ is 2. Therefore $(0, 2)$ is a point on the graph of the equation.

(2)  The slope of $y = \dfrac{-3}{4}x + 2$ is $\dfrac{-3}{4}$. Therefore as the value of $x$ is increased by 1 then the value of $y$ must decrease $\left(\dfrac{-3}{4}\text{ is negative}\right)$ by $\dfrac{3}{4}$.

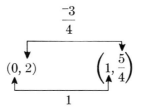

Hence $\left(1, \dfrac{5}{4}\right)$ is a second solution of $y = \dfrac{-3}{4}x + 2$ and the graph is shown in figure 5.8.

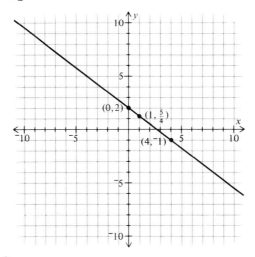

FIGURE 5.8

Notice that the graph of $y = \dfrac{-3}{4}x + 2$ in figure 5.8 passes through the point representing $(4, {}^-1)$. This ordered pair could have been found using the slope $\dfrac{-3}{4}$ because as the value of $x$ increases by 4 then the value of $y$ must be altered by $\dfrac{-3}{4} \cdot 4$ or $^-3$.

$$
\begin{array}{ccc}
& {}^-3 & \\
(0, 2) & & (4, {}^-1) \\
& 4 &
\end{array}
$$

In graphing an equation of the form $y = mx + b$, the $y$ intercept, $b$, is used to find the solution on the $y$ axis. The slope, $m$, is used to find a second solution of the equation and if $m$ is a rational number, $\dfrac{c}{d}$ , the numerator, $c$, can be considered the change in the value of $y$ whenever the denominator, $d$, is the change in the value of $x$.

For the equation $y = \dfrac{2}{3}x - 5$ the $y$ intercept is $^-5$ and the solution on the $y$ axis is $(0, {}^-5)$. The slope of $y = \dfrac{2}{3}x - 5$ is $\dfrac{2}{3}$ which means that when the value of $x$ is increased by 3 then the value of $y$ must increase by 2. Hence $(3, {}^-3)$ is also a solution of $y = \dfrac{2}{3}x - 5$.

$$
\begin{array}{ccc}
& 2 & \\
(0, {}^-5) & & (3, {}^-3) \\
& 3 &
\end{array}
$$

The graph of $y = \dfrac{2}{3}x - 5$ using the solutions $(0, {}^-5)$ and $(3, {}^-3)$ is shown in figure 5.9.

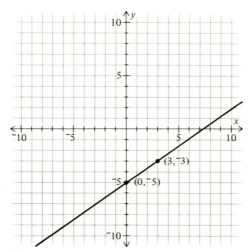

FIGURE 5.9

EXERCISE 5.11.3

Graph each equation using the $y$ intercept and slope.

1.  $y = 2x - 3$

2.  $y = {}^-4x + 5$

3.  $y = x + 1$

4.  $y = {}^-3x - 2$

5.  $y = {}^-x + 6$

6.  $y = 5x + 6$

7.  $y = {}^-2x - 4$

8.  $y = 4x - 4$

9.  $y = 2x + 7$

10.  $y = {}^-3x + 6$

11.  $y = \dfrac{4}{5} x - 3$

12.  $y = \dfrac{{}^-2}{5} x - 4$

13.  $y = \dfrac{3}{2} x + 4$

14.  $y = \dfrac{5}{3} x + 2$

15.  $y = \dfrac{{}^-1}{4} x - 1$

16.  $y = \dfrac{{}^-3}{4} x - 5$

17.  $y = {}^-4x + 10$

18.  $y = \dfrac{2}{3} x$

19.  $y = {}^-x - 7$

20.  $y = 3x - 4$

### 5.12   Graphing Inequalities

$2x - 3y > 6$ is an inequality in two variables. $(5, 1)$ is a solution for $2x - 3y > 6$ because $2 \cdot 5 - 3 \cdot 1 > 6$ is a true statement. $(4, 2)$ is not a solution for $2x - 3y > 6$ because $2 \cdot 4 - 3 \cdot 2 > 6$ is a false statement. To graph $2x - 3y > 6$ the following steps are used:

(1)   Graph the equation $2x - 3y = 6$.

(2)   Choose any ordered pair that is not a solution of $2x - 3y = 6$ and therefore is not on the line of the equation.

(3)   If the ordered pair chosen in step 2 makes $2x - 3y > 6$ a true statement then the graph consists of all points on the same side of $2x - 3y = 6$ as the chosen ordered pair.

(4)   If the ordered pair chosen in step 2 makes $2x - 3y > 6$ a false statement then the graph consists of all points on the opposite side of $2x - 3y = 6$ as the chosen ordered pair.

The graphing of $2x - 3y > 6$ is accomplished in the following steps:

(1)   Graph $2x - 3y = 6$.

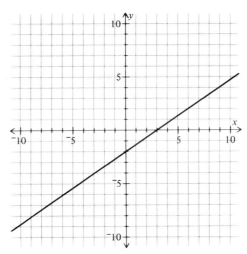

FIGURE 5.10

(2)   Choose any ordered pair that is not a solution of $2x - 3y = 6$. $(0, 0)$ is usually an easy ordered pair to use. $2 \cdot 0 - 3 \cdot 0 > 6$ is false.

(3) Since $(0, 0)$ is not a solution of $2x - 3y > 6$, the graph consists of all those points below the line of $2x - 3y = 6$.

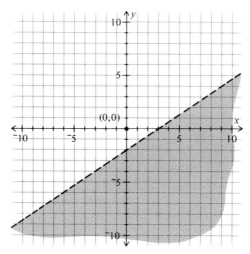

FIGURE 5.11

The graphing of $y > 2x - 3$ is accomplished in the following steps:

(1) Graph $y = 2x - 3$.

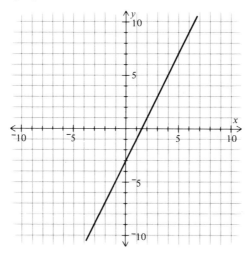

FIGURE 5.12

(2) Choose any ordered pair that is not a solution of $y = 2x - 3$. $(0, 0)$ is used here and $0 > 2 \cdot 0 - 3$ is true.

(3)   The graph of $y > 2x - 3$ consists of all those points on the same side of the line as $(0, 0)$.

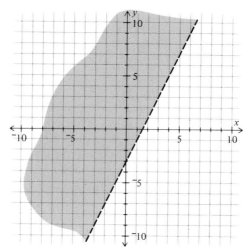

FIGURE 5.13

EXERCISE 5.12.1

Graph the following inequalities:

1.  $x + y > 4$                  11.  $y > 3x + 1$
2.  $x + 2y < 8$                 12.  $y < {}^-2x - 3$

3.  $x - 3y > 6$                 13.  $y < \dfrac{2}{3}x + 2$

4.  $2x + y < 5$                 14.  $y > \dfrac{{}^-3}{4}x - 5$

5.  $4x - y > 2$                 15.  $y > \dfrac{{}^-1}{2}x + 4$

6.  $2x - y < 7$                 16.  $y > 2x + 7$
7.  $x + 3y > {}^-2$             17.  $2x + 3y < 9$

8.  $2x + 5y > 3$                18.  $y < \dfrac{{}^-3}{4}x - 2$

9.  $3x + y < {}^-2$             19.  $y > \dfrac{5}{6}x + 3$

10.  $x - y > 2$                 20.  $y < {}^-3x + 5$

## 5.13   Graphing Pairs of Equations

The equations $5x - 3y = 2$ and $x + y = 10$ both have a graph that is a straight line. Each straight line represents ordered pairs that are solutions for its equation. Those points that are not on the line of an equation are not solutions of the equation.

An interesting result is obtained by graphing both $5x - 3y = 2$ and $x + y = 10$ on the same pair of axes.

Figure 5.14 shows the straight line graphs of $5x - 3y = 2$ and $x + y = 10$. Notice that the two lines have exactly one point in common, $(4, 6)$. Although each equation has no limit to its number of ordered pair

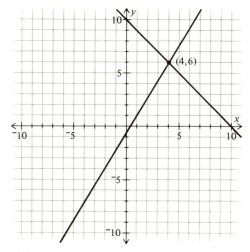

FIGURE 5.14

solutions, the graph of figure 5.14 shows clearly that there is only one ordered pair that is shared by both equations as a solution. We say that $(4, 6)$ is the *common solution* for $5x - 3y = 2$ and $x + y = 10$.

EXERCISE 5.13.1

Graph each pair of equations on a single set of axes and find their common solution by finding the point that is on both lines.

1.  $x - y = 3$   and   $2x + y = 12$
2.  $4x + 3y = 0$   and   $x + y = 1$
3.  $2x - 5y = 3$   and   $x - 2y = 2$
4.  $4x - 3y = 8$   and   $x - y = 2$
5.  $x + 3y = 1$   and   $x - y = 9$

6. $x + 2y = {}^-1$   and   $2x + y = 7$
7. $x + y = 0$   and   $x - 3y = 4$
8. $x - 3y = 4$   and   $x + y = {}^-8$
9. $x - 2y = {}^-8$   and   $x + 4y = {}^-2$
10. $x - y = 3$   and   $x - 2y = 6$

## 5.14  Finding Common Solutions by the Addition Method

In the last exercise common solutions for pairs of equations were found by graphing. In this section another method for finding common solutions is explained.

To solve $2x - 19 = 47$ the first step is the addition of 19 to both sides of the equation. This step is justified because if the same quantity is added to both members of an equality then a new equivalent equality is obtained. The same type of justification is involved in finding the common solution for $2x + y = 14$ and $x - y = 1$.

To solve $2x + y = 14$ and $x - y = 1$ for their common solution we first note that the equation $x - y = 1$ makes the claim that $(x - y)$ is equal to 1. This being the case, $(x - y)$ may be added to one side of $2x + y = 14$ and 1 added to the other side of the equation to obtain a new equation whose truth set will lead to the common solution for the pair of equations.

$$2x + y = 14 \qquad x - y = 1$$

$$(2x + y) + (x - y) = 14 + 1$$

$$3x = 15$$

$$x = 5$$

If $x$ is replaced by 5 in either $2x + y = 14$ or $x - y = 1$ then the corresponding value of $y$ for the common solution of the two equations will be found.

$$x - y = 1 \qquad x = 5$$

$$5 - y = 1$$

$$-y = {}^-4$$

$$y = 4$$

$(5, 4)$ is the common solution for $2x + y = 14$ and $x - y = 1$. Because of the manner in which $(5, 4)$ was acquired it is probably clear that it is a solution for $x - y = 1$, but the student should check for himself to see that it is also a solution of $2x + y = 14$.

As another example of finding the common solution by addition consider the following pair of equations:

$$x - 2y = 7 \qquad {}^-x + 4y = 6$$

$$(x - 2y) + ({}^-x + 4y) = 7 + 6$$

$$2y = 13$$

$$y = \frac{13}{2}$$

$$x - 2y = 7 \quad \text{and} \quad y = \frac{13}{2}$$

$$x - 2 \cdot \frac{13}{2} = 7$$

$$x - 13 = 7$$

$$x = 20$$

$\left(20, \dfrac{13}{2}\right)$ is the common solution for $x - 2y = 7$ and ${}^-x + 4y = 6$.


EXERCISE 5.14.1

Find common solutions by addition for each pair of equations.

1.   $3x + y = 2$   and   $x - y = 6$
2.   $4x - y = 9$   and   ${}^-x + y = 6$
3.   $x + 3y = 2$   and   $2x - 3y = {}^-8$
4.   $2x - 2y = 7$   and   $x + 2y = 2$
5.   $x - 2y = 5$   and   ${}^-x - 3y = 10$
6.   ${}^-x + y = 4$   and   $x + 5y = 8$
7.   $3x - 2y = 9$   and   ${}^-3x - 2y = 7$
8.   ${}^-2x + 3y = 1$   and   $2x - y = 3$
9.   $2x + y = 7$   and   $x - y = 1$
10.   $4x - 3y = 2$   and   $x + 3y = 8$
11.   $2x + y = 9$   and   $3x - y = 16$
12.   $4x - 3y = 7$   and   $2x + 3y = 2$
13.   $2x + 7y = 6$   and   ${}^-2x + y = 2$
14.   $5x + 2y = 0$   and   $x - 2y = 6$
15.   $x - 5y = 6$   and   ${}^-x + y = 4$

16. $2x + 7y = 1$   and   $2x - 7y = 3$
17. $^-3x - y = 5$   and   $3x + 8y = ^-5$
18. $x + 4y = 9$   and   $3x - 4y = ^-1$
19. $5x - y = 6$   and   $^-5x - y = 10$
20. $x + 5y = 9$   and   $3x - 5y = ^-5$

The method of finding common solutions used in the last exercise needs a slight refinement to be applied to the equations $5x - 2y = 11$ and $x - y = 4$.

Notice that if the two members of the equations $5x - 2y = 11$ and $x - y = 4$ are added then neither variable is eliminated and therefore little, if any, progress has been made in finding the common solution.

$$5x - 2y = 11 \qquad x - y = 4$$

$$(5x - 2y) + (x - y) = 11 + 4$$

$$6x - 3y = 15$$

One of the variables should be eliminated to accomplish the task of finding a common solution. Neither variable is eliminated by adding $5x - 2y = 11$ and $x - y = 4$ because the coefficients of neither $x$ nor $y$ are opposites.

The opposite of $^-2y$ is $2y$. The variable $y$ could be eliminated from $5x - 2y = 11$ and $x - y = 4$ if the coefficient of $y$ in the second equation were 2. This change of coefficients can be accomplished using the fact that any nonzero number may be multiplied by both members of an equation to provide a new, equivalent equation.

To change the coefficient of $y$ in $x - y = 4$ to 2 we multiply both members of the equation by $^-2$ because $^-2 \cdot {}^-1 = 2$.

$$x - y = 4$$

$$^-2(x - y) = {}^-2 \cdot 4$$

$$^-2x + 2y = {}^-8$$

The steps in finding the common solution for $5x - 2y = 11$ and $x - y = 4$ are

(1)   Since $^-2y$ and $2y$ are opposites, multiply both members of $x - y = 4$ by $^-2$.

$$^-2(x - y) = {}^-2 \cdot 4$$

$$^-2x + 2y = {}^-8$$

(2)   $x - y = 4$ and $^-2x + 2y = ^-8$ are equivalent equations.

Add $5x - 2y = 11$ and $^-2x + 2y = ^-8$ to eliminate $y$.

$$(5x - 2y) + (^-2x + 2y) = 11 + ^-8$$

$$3x = 3$$

$$x = 1$$

(3)   Complete the ordered pair solution for $5x - 2y = 11$ and $x - y = 4$ by substituting 1 for $x$ in either of the equations.

$$5x - 2y = 11 \qquad x = 1$$

$$5 - 2y = 11$$

$$^-2y = 6$$

$$y = ^-3$$

$(1, ^-3)$ is the common solution for $5x - 2y = 11$ and $x - y = 4$.

EXERCISE 5.14.2

Find common solutions:

1.  $3x - y = 7$   and   $2x + 3y = 12$   (Multiply both members of $3x - y = 7$ by 3.)
2.  $2x - 5y = ^-12$   and   $x + 2y = 3$   (Multiply both members of $x + 2y = 3$ by $^-2$ to eliminate $x$.)
3.  $x + y = 2$   and   $3x + y = 12$   (Multiply both members of one of the equations by $^-1$.)
4.  $3x - 2y = 5$   and   $5x + 4y = 12$   (Multiply both members of $3x - 2y = 5$ by 2.)
5.  $x + 3y = 7$   and   $x - 4y = ^-14$   (Multiply both members of one of the equations by $^-1$.)
6.  $2x + y = 9$   and   $x + y = 5$
7.  $x - 4y = 5$   and   $x - 3y = 4$
8.  $3x - 2y = 8$   and   $2x + y = 17$
9.  $4x + 3y = ^-4$   and   $2x - y = 8$
10.  $2x + 5y = 6$   and   $4x - 3y = ^-1$
11.  $2x + y = 7$   and   $x - 6y = ^-29$
12.  $5x - 2y = 14$   and   $3x + 4y = 24$
13.  $x + 7y = 5$   and   $2x - 3y = 10$
14.  $2x - 3y = 9$   and   $3x - y = 17$

15. $x + 4y = {}^-6$ and $5x - 7y = {}^-3$
16. $2x - y = 7$ and $3x + 2y = 21$
17. $x + 6y = 9$ and $2x - 4y = 2$
18. $5x + 4y = 2$ and $3x - 2y = {}^-1$
19. $3x - 2y = 7$ and $x + 5y = {}^-9$
20. $2x - y = 5$ and $7x - 2y = {}^-21$

The procedure of the last exercise for finding common solutions leads directly to a method for handling $3x - 2y = 7$ and $2x - 5y = {}^-10$. For these two equations the common solution will be found by multiplying each of the equations by suitably selected numbers.

To eliminate the $x$'s from $3x - 2y = 7$ and $2x - 5y = {}^-10$ we notice that 6 is the least integer that is a common multiple for 2 and 3, the coefficients of $x$. The common solution for $3x - 2y = 7$ and $2x - 5y = {}^-10$ is found in the following steps:

(1)  Multiply $3x - 2y = 7$ by 2 and $2x - 5y = {}^-10$ by $^-3$.

$$2(3x - 2y) = 2 \cdot 7 \qquad\qquad 6x - 4y = 14$$

$$^-3(2x - 5y) = {}^-3 \cdot {}^-10 \qquad {}^-6x + 15y = 30$$

(2)  Add $6x - 4y = 14$ and $^-6x + 15y = 30$.

$$(6x - 4y) + ({}^-6x + 15y) = 14 + 30$$

$$11y = 44$$

$$y = 4$$

(3)  Since $y = 4$, use either $3x - 2y = 7$ or $2x - 5y = {}^-10$ to find the replacement for $x$ in the common solution.

$$3x - 2y = 7 \qquad y = 4$$

$$3x - 8 = 7$$

$$3x = 15$$

$$x = 5$$

$(5, 4)$ is the common solution for $3x - 2y = 7$ and $2x - 5y = {}^-10$.

For the equations $4x - 7y = 1$ and $6x - 5y = {}^-15$ either variable, $x$ or $y$, may be eliminated. To eliminate the $x$'s it would be necessary to find a common multiple for 4 and 6 since they are the coefficients of the $x$'s. To eliminate the $y$'s it is necessary to find a common multiple for $^-7$ and $^-5$ since they are the coefficients of the $y$'s.

The following steps show how to find the common solution for $4x - 7y = 1$ and $6x - 5y = {}^-15$ by first eliminating the $y$'s.

(1)   Since the $y$ coefficients in $4x - 7y = 1$ and $6x - 5y = {}^-15$ are $^-7$ and $^-5$, the first step is to find a common multiple for $^-7$ and $^-5$.

> 35 is a common multiple for $^-7$ and $^-5$

(2)   Since 35 is a common multiple for $^-7$ and $^-5$, we multiply each member of $4x - 7y = 1$ by 5 and each member of $6x - 5y = {}^-15$ by $^-7$.

$$5(4x - 7y) = 5 \cdot 1 \qquad 20x - 35y = 5$$

$$^-7(6x - 5y) = {}^-7 \cdot {}^-15 \qquad {}^-42x + 35y = 105$$

(3)   Adding the equations obtained in step 2, the $y$'s are eliminated.

$$(20x - 35y) + ({}^-42x + 35y) = 5 + 105$$

$$^-22x = 110$$

$$x = {}^-5$$

(4)   Since $x = {}^-5$, use either $4x - 7y = 1$ or $6x - 5y = {}^-15$ to find the corresponding value for $y$ in the common solution.

$$4x - 7y = 1 \qquad x = {}^-5$$

$$4 \cdot {}^-5 - 7y = 1$$

$$^-7y = 21$$

$$y = {}^-3$$

(5)   $({}^-5, {}^-3)$ is the common solution for $4x - 7y = 1$ and $6x - 5y = {}^-15$.

## EXERCISE 5.14.3

Find common solutions:

1. $2x - 3y = 7$   and   $5x + 2y = 27$   (Multiply the first equation by 2 and the second by 3 to eliminate the $y$'s.)
2. $5x + 8y = 11$   and   $3x - 5y = {}^-13$   (Multiply the first equation by 3 and the second by $^-5$ to eliminate the $x$'s.)
3. $2x - 3y = 19$   and   $5x - 7y = 45$   (Multiply the first by $^-7$ and the second by 3.)

4. $4x - 5y = 20$  and  $3x - 2y = 10$
5. $4x - 3y = 7$  and  $5x + 2y = 3$
6. $3x - 2y = {}^-16$  and  $5x - 7y = {}^-45$
7. $5x - 3y = {}^-7$  and  $4x - 11y = 3$
8. $3x + 5y = {}^-9$  and  $7x - 6y = 45$
9. $4x - 9y = 19$  and  $2x + 3y = {}^-13$
10. $2x - 13y = {}^-5$  and  $9x - 5y = 31$
11. $2x + 5y = 3$  and  $3x - 2y = {}^-24$
12. $5x + 2y = 15$  and  $2x - 3y = {}^-13$
13. $3x + 2y = 14$  and  $7x + 5y = {}^-35$
14. $6x - 5y = 2$  and  $4x - 3y = 0$
15. $7x - 2y = 10$  and  $5x - 5y = 0$
16. $2x + 3y = {}^-9$  and  $3x - 5y = 34$
17. $2x + 3y = 14$  and  $5x + 4y = 0$
18. $7x - 5y = {}^-21$  and  $3x + 8y = {}^-9$
19. $4x + 3y = 13$  and  $9x + 8y = 23$
20. $5x + 3y = 2$  and  $7x + 5y = {}^-2$

Not every pair of equations in two variables has a common solution. Shown in figure 5.15 is a graph with the lines of $2x - 3y = 6$ and $2x - 3y = {}^-12$.

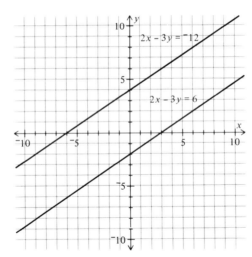

FIGURE 5.15

Notice that the lines are parallel and therefore never intersect. Hence there is no solution shared by the two equations and graphically we can see that there is no common solution for $2x - 3y = 6$ and $2x - 3y = {}^-12$.

The fact that $2x - 3y = 6$ and $2x - 3y = {}^-12$ have no common solution need not be discovered by graphing the lines of the equations. It is also obvious when the addition method for finding common solutions is attempted.

To eliminate the $x$'s in $2x - 3y = 6$ and $2x - 3y = {}^-12$, one of the equations should be multiplied by $^-1$.

$$^-1(2x - 3y) = {}^-1 \cdot 6 \qquad {}^-2x + 3y = {}^-6$$

Now adding $^-2x + 3y = {}^-6$ and $2x - 3y = {}^-12$ we obtain

$$({}^-2x + 3y) + (2x - 3y) = {}^-6 + {}^-12$$

$$0 = {}^-18$$

Since $0 = {}^-18$ is false, we can conclude that there are no values for $x$ and $y$ that will serve as a common solution for $2x - 3y = 6$ and $2x - 3y = {}^-12$.

As another example of a pair of equations having no common solution let us attempt to solve $x + 2y = 9$ and $3x + 6y = 1$ by addition.

To eliminate the $x$'s from $x + 2y = 9$ and $3x + 6y = 1$ we can multiply the first equation by $^-3$.

$$^-3(x + 2y) = {}^-3 \cdot 9 \qquad {}^-3x - 6y = {}^-27$$

Adding $^-3x - 6y = {}^-27$ and $3x + 6y = 1$ we obtain

$$({}^-3x - 6y) + (3x + 6y) = {}^-27 + 1$$

$$0 = {}^-26$$

$0 = {}^-26$ is a false statement. Therefore there is no common solution for $x + 2y = 9$ and $3x + 6y = 1$.

EXERCISE 5.14.4

Find the common solution or show that there is no common solution.

1.  $4x - 3y = 7$  and  $4x - 3y = {}^-4$
2.  $2x + y = 11$  and  $6x + 3y = 2$
3.  $x - 2y = 5$  and  $x + y = 11$
4.  $4x - y = 11$  and  $3x - 4y = 5$
5.  $3x - 6y = 1$  and  $4x - 8y = {}^-3$
6.  $x + 3y = 4$  and  $x - 2y = {}^-11$
7.  $5x - 15y = 4$  and  $x - 3y = 7$
8.  $4x - 6y = 3$  and  $6x - 9y = 1$
9.  $3x - 2y = 6$  and  $2x - 3y = {}^-1$
10.  $5x - y = 4$  and  $x + 5y = 6$

### 5.15  Finding Common Solutions by the Substitution Method

In this chapter two methods, graphing and addition, for finding common solutions for pairs of linear equations have been studied. There is another method, called substitution, which is sometimes used.

In a football or basketball game when a new player enters the contest he replaces one of the other players. We call the new player a substitute. In Algebra we may find it desirable to replace a variable by some other algebraic expression. Whenever this is done we call it substitution.

If we are given the equation $x = 5y$ then we can replace the letter $x$ by the expression $5y$ because any pair of values $(x, y)$ that makes $x = 5y$ a true equality will make $5y$ an equal substitution for $x$. If we were given the expression $2x + 3y$ along with the equation $x = 5y$ then $5y$ can be substituted for $x$ in the expression $2x + 3y$ as shown below.

$$2x + 3y \qquad x = 5y$$

$$2(5y) + 3y$$

$$10y + 3y$$

$$13y$$

If we are given the expression $4x - 3y$ and the equation $y = 2x - 5$ then the expression $(2x - 5)$ can be substituted for $y$ because that is the claim of the equality. Replacing $y$ by $(2x - 5)$ in the expression $4x - 3y$ results in the following simplification.

$$4x - 3y \qquad y = 2x - 5$$

$$4x - 3(2x - 5)$$

$$4x - 6x + 15$$

$$^-2x + 15$$

EXERCISE 5.15.1

In each problem an equation and an algebraic expression are given. Use the equation and determine a substitution for one of the variables in the algebraic expression. After the substitution simplify the result.

1.  $x = 2y$  and  $4x - 3y$
2.  $y = 3x$  and  $2x - 5y$
3.  $x = ^-5y$  and  $4x + y$
4.  $x = ^-4y$  and  $x + 7y$
5.  $y = 8x$  and  $5x - 6y$
6.  $y = ^-x$  and  $3x - 2y$
7.  $x = 3y + 2$  and  $2x - 5y$

8.  $x = 5 - 3y$   and   $3x + 5y$
9.  $y = 4 - x$   and   $x - 4y$
10.  $y = 8x + 3$   and   $3x - y$
11.  $x = 6y$   and   $5x - 7y$
12.  $y = {}^-3x$   and   $4x - y$
13.  $x = 4y$   and   $3x + 6y$
14.  $x = {}^-y$   and   $2x + y$
15.  $y = 7x$   and   $x - 2y$
16.  $y = {}^-5x$   and   $3x + y$
17.  $x = 2y - 3$   and   $4x - 5y$
18.  $y = x + 5$   and   $2x + 3y$
19.  $x = 3y + 7$   and   $3x - 4y$
20.  $y = 3x - 1$   and   $2x - 5y$

To find the common solution for $3x - 2y = 5$ and $y = 4x$ the substitution method may be used by replacing $y$ in the first equation by its matching value from the second equation, $4x$.

$$3x - 2y = 5 \qquad y = 4x$$
$$3x - 2(4x) = 5$$
$$3x - 8x = 5$$
$${}^-5x = 5$$
$$x = {}^-1$$
$$y = 4x \quad \text{and} \quad x = {}^-1$$
$$y = 4 \cdot {}^-1$$
$$y = {}^-4$$

Hence $({}^-1, {}^-4)$ is the common solution for $3x - 2y = 5$ and $y = 4x$. As a second example of the substitution method the common solution of $x = 3y - 1$ and $2x - y = 8$ is shown below.

$$x = 3y - 1 \qquad\qquad 2x - y = 8$$
$$2(3y - 1) - y = 8$$
$$6y - 2 - y = 8$$
$$5y = 10$$
$$y = 2$$
$$x = 3y - 1 \quad \text{and} \quad y = 2$$
$$x = 3 \cdot 2 - 1$$
$$x = 5$$

Hence $(5, 2)$ is the common solution for $x = 3y - 1$ and $2x - y = 8$.

Whenever a pair of equations has no common solution the substitution method leads to a numerical statement that is false. For example, if the substitution method is applied to $3x - y = 7$ and $y = 3x$ the following steps might be used.

$$3x - y = 7 \qquad y = 3x$$

$$3x - (3x) = 7$$

$$0 = 7$$

Since $0 = 7$ is a false numerical statement we conclude that there is no common solution for $3x - y = 7$ and $y = 3x$.

EXERCISE 5.15.2

Use the substitution method to find the common solution or to show that the pair of equations has no common solution.

1. $5x - y = 8$   and   $y = 3x$
2. $x - y = 8$   and   $y = 5x$
3. $x - 5y = {}^{-}14$   and   $x = 3y$
4. $2x - 4y = 7$   and   $x = 2y$
5. $x + y = 7$   and   $y = 2x + 1$
6. $4x + 3y = 9$   and   $y = 4 - x$
7. $3x - 2y = 14$   and   $x = 2y + 6$
8. $3x + y = 2$   and   $y = {}^{-}5x$
9. $6x - 3y = 7$   and   $y = 2x - 3$
10. $2x + y = 11$   and   $x = 2y - 7$
11. $y = 2x$   and   $x + y = 9$
12. $x = {}^{-}5y$   and   $2x + y = 9$
13. $x = 2y - 1$   and   $3x - 4y = 3$
14. $2x + y = 5$   and   $y = {}^{-}x$
15. $3x + 4y = 4$   and   $y = 3x + 1$
16. $y = 4x - 5$   and   $x - y = 2$
17. $2x + 5y = 48$   and   $y = 2x$
18. $x = 5 - 2y$   and   $2x + y = 1$
19. $x = {}^{-}4y$   and   $3x - 5y = 17$
20. $2x - 3y = 6$   and   $x = 2y + 2$

Three methods for finding common solutions of pairs of linear equations have been studied in this chapter. They are graphing, addition, and substitution. Any one of the three methods could be used on a pair of linear equations, however the graphing method has obvious handicaps because the procedure is often time consuming and even then only supplies approximate solutions when the results are not integers. In general, therefore, the addition and substitution methods are most commonly used in finding common solutions. Either method will lead to the common solution (or no solution) for any pair of linear equations. Therefore, the student may choose either method without fearing that it is the "wrong" approach to the problem.

The reason both the addition and substitution methods are taught in this Algebra text is that in some instances one of the procedures may be easier than the other.

For example, to find the common solution for $2x - 3y = 7$ and $x + 3y = 8$ the addition method is quite easy to apply.

$$2x - 3y = 7$$
$$\underline{x + 3y = 8}$$
$$3x = 15$$
$$x = 5$$

Once we know $x = 5$, it is fairly easy to complete the common solution, $(5, 1)$. The addition method is certainly easy for the equations $2x - 3y = 7$ and $x + 3y = 8$ because one of the variables was quickly and easily eliminated.

As an example in which the substitution method is probably easier, consider the following pair of equations: $5x - 2y = {}^-2$ and $y = 3x$. Since the second equation is already solved for $y$ then $3x$ can be substituted for $y$ in the first equation.

$$5x - 2y = {}^-2 \qquad y = 3x$$
$$5x - 2(3x) = {}^-2$$
$$5x - 6x = {}^-2$$
$${}^-x = {}^-2$$
$$x = 2$$

The common solution $(2, 6)$ follows immediately.

In the following exercise you are asked to find common solutions for pairs of linear equations. Use either the addition or substitution method.

EXERCISE 5.15.3

Find the common solution or state that there is no common solution.

1.  $2x + y = 15, \quad x - y = 6$
2.  $3x + 4y = {}^-18, \quad y = {}^-3x$
3.  $4x - 7y = 5, \quad x = 2y$

4. $3x + 5y = 7$, $x + 2y = 4$
5. $x + y = 1$, $4x - 3y = {}^-24$
6. $4x - y = 21$, $x = y$
7. $x + 5y = 11$, $y = x + 13$
8. $4x - y = 1$, $y = 4x + 7$
9. $5x - 3y = {}^-15$, $y = x + 3$
10. $2x + 7y = 31$, $5x - 3y = {}^-25$
11. $5x - 3y = 4$, $10x - 6y = 5$
12. $7x - y = 4$, $y = 3x$
13. $3x - y = 2$, $x + y = 4$
14. $x + 4y = 17$, $x - 3y = {}^-11$
15. $4x = y$, $x - y = 6$
16. $x = 3y - 1$, $x - 7y = 3$
17. $2x - 3y = 4$, $2x + 5y = 20$
18. $x - 5y = 10$, $x = {}^-5y$
19. $x + 3y = 0$, $7x + 20y = 5$
20. $y = {}^-2x$, $5x + 3y = 6$

### 5.16   Set Notation for Sets of Ordered Pairs

An open sentence in two variables such as $2x + y = 7$ has no limit to its number of solutions. Graphically we show the solutions of $2x + y = 7$ by drawing its line. It would be impossible to list all the solutions of $2x + y = 7$ in braces, but we indicate the set of all solutions for the equation by the following notation.

$$\{(x, y) \mid 2x + y = 7\}$$

The braces indicate a set and, reading from left to right, the ordered pair $(x, y)$ indicates that each element of the set is to be an ordered pair. The vertical bar to the right of $(x, y)$ is to be read "such that" and the equation, $2x + y = 7$, is used to test ordered pairs for possible membership in the set.

A word translation of "$\{(x, y) \mid 2x + y = 7\}$" would read as "the set of all ordered pairs $(x, y)$ such that $2x + y = 7$ becomes a true numerical statement."

$(4, {}^-1)$ is an element of $\{(x, y) \mid 2x + y = 7\}$ because $(4, {}^-1)$ is an ordered pair that is a solution of $2x + y = 7$.

$(5, 2)$ is not an element of $\{(x, y) \mid 2x + y = 7\}$ because $(5, 2)$ is not a solution of $2x + y = 7$.

7 is not an element of $\{(x, y) \mid 2x + y = 7\}$ simply because every element of the set is an ordered pair and 7 is not an ordered pair.

"$\in$" is the symbol for "is an element of." "$\notin$" is the symbol for "is not an element of."

EXERCISE 5.16.1

Determine the truth or falsity of each of the following.

1. $(7, 1) \in \{(x, y) \mid x - 3y = 4\}$
2. $(6, ^-3) \in \{(x, y) \mid 2x + y = 15\}$
3. $(^-5, ^-1) \in \{(x, y) \mid x - 3y = ^-2\}$
4. $(7, ^-2) \notin \{(x, y) \mid 2x + y = 16\}$
5. $(5, ^-4) \in \{(x, y) \mid x + 3y = ^-7\}$
6. $(^-2, 1) \in \{(x, y) \mid 3x + y = 5\}$
7. $6 \in \{(x, y) \mid x + y = 6\}$
8. $4 \in \{(x, y) \mid x - 3y = 4\}$
9. $(1, 3) \in \{(x, y) \mid x + y > 2\}$
10. $(^-2, 5) \in \{(x, y) \mid x + y > 2\}$
11. $(^-3, 6) \notin \{(x, y) \mid x + y > 2\}$
12. $(4, 1) \notin \{(x, y) \mid x - y < 3\}$
13. $(6, ^-2) \in \{(x, y) \mid 3x - 5y < 20\}$
14. $(4, 7) \in \{(x, y) \mid 5x - 2y = 6\}$
15. $(^-2, ^-1) \in \{(x, y) \mid x - 2y = 0\}$
16. $(5, ^-3) \notin \{(x, y) \mid y = x - 8\}$
17. $(2, 7) \in \{(x, y) \mid y < 3x - 4\}$
18. $(^-4, ^-3) \notin \{(x, y) \mid x + y > ^-2\}$
19. $^-5 \in \{(x, y) \mid x + y < 0\}$
20. $(5, ^-1) \in \{(x, y) \mid 3x - y = 16\}$

## 5.17 Intersecting Sets of Ordered Pairs

When we speak of the common solution for two linear equations we mean the ordered pair that is a solution of both equations. Recall from chapter 1 that when we were interested in finding those elements that two sets had in common then we used the set operation of intersection. To intersect two sets is to form a set containing all those elements which were in both the original sets.

Since $(5, 2)$ is the common solution of $2x - y = 8$ and $x + y = 7$ we can write this in set notation as

$$\{(x, y) \mid 2x - y = 8\} \cap \{(x, y) \mid x + y = 7\} = \{(5, 2)\}$$

Though each of the sets, $\{(x, y) \mid 2x - y = 8\}$ and $\{(x, y) \mid x + y = 7\}$, has an unlimited number of elements, the only ordered pair that belongs to both sets is $(5, 2)$. Hence the intersection set is the singleton set $\{(5, 2)\}$.

Graphically the intersection of the two sets $\{(x, y) \mid 2x - y = 8\}$ and $\{(x, y) \mid x + y = 7\}$ is shown by finding all points that are on both lines at the same time. In figure 5.16 the two lines are shown. Notice that the only point common to the two lines is (5, 2).

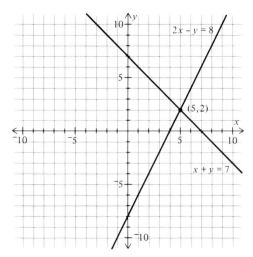

FIGURE 5.16

EXERCISE 5.17.1

Find the intersection of each of the following sets by the graphing method.

1. $\{(x, y) \mid x + y = 5\} \cap \{(x, y) \mid 3x + y = 7\}$
2. $\{(x, y) \mid x - y = 3\} \cap \{(x, y) \mid 2x + y = 3\}$
3. $\{(x, y) \mid x + 2y = 11\} \cap \{(x, y) \mid x - 2y = ^-1\}$
4. $\{(x, y) \mid 3x - y = 5\} \cap \{(x, y) \mid x + 2y = 18\}$
5. $\{(x, y) \mid x + 3y = 10\} \cap \{(x, y) \mid 5x + 4y = 6\}$
6. $\{(x, y) \mid x + y = 0\} \cap \{(x, y) \mid 4x + 3y = 5\}$
7. $\{(x, y) \mid x + y = 6\} \cap \{(x, y) \mid x + y = 2\}$
8. $\{(x, y) \mid 2x - y = 5\} \cap \{(x, y) \mid 2x - y = 3\}$
9. $\{(x, y) \mid x + y = 4\} \cap \{(x, y) \mid 2x + 2y = 8\}$
10. $\{(x, y) \mid x - 3y = 6\} \cap \{(x, y) \mid 2x - 6y = 12\}$

Problems 7 through 10 of the last exercise provide interesting examples for your intuitive understanding of intersection as well as your earlier acquired skills at finding common solutions by addition.

In problems 7 and 8 you should have found that the lines were parallel. Hence they have no points in common and the intersection of the two sets has no elements. Therefore, it is the empty set, { }.

In problems 9 and 10 you should have found that the lines were identical. Hence every point of one line was also a point on the other. Therefore, the intersection consists of all points on the common line.

The intersection of $\{(x, y) \mid 2x - y = 7\}$ and $\{(x, y) \mid x + y < 6\}$ is slightly different from the problems of the last exercise because one of the sets involves an inequality. Graphing both sets on the same pair of axes results in figure 5.17. Since $\{(x, y) \mid 2x - y = 7\} \cap \{(x, y) \mid x + y < 6\}$ is

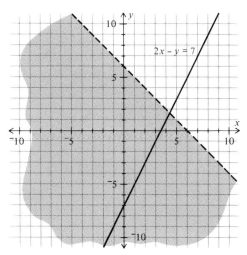

FIGURE 5.17

to consist of all points common to *both* sets, it must contain points on the line of $2x - y = 7$ *and* to the lower left of the line of $x + y = 6$. Hence the intersection is the half-line shown in figure 5.18.

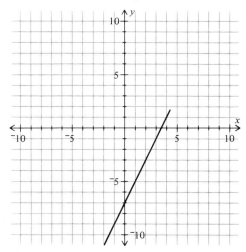

FIGURE 5.18

For the intersection of $\{(x, y) \mid x - y < 4\}$ and $\{(x, y) \mid 2x - 3y > 6\}$ the two sets should be graphed on the same pair of axes as shown in figure 5.19.

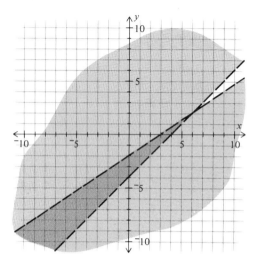

FIGURE 5.19

Since $\{(x, y) \mid x - y < 4\} \cap \{(x, y) \mid 2x - 3y > 6\}$ is the set of all points common to both sets it consists of all those points in the sector marked twice in figure 5.19. The intersection of the two sets is shown graphically in figure 5.20.

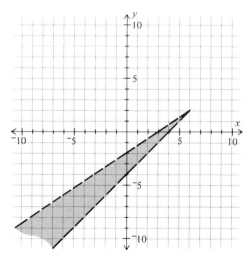

FIGURE 5.20

EXERCISE 5.17.2

Find the intersection sets by graphing.

1.  $\{(x, y) \mid x + y > 1\} \cap \{(x, y) \mid 2x - y = 4\}$
2.  $\{(x, y) \mid 3x + 2y = 4\} \cap \{(x, y) \mid x + 2y = 6\}$
3.  $\{(x, y) \mid 2x - y < 6\} \cap \{(x, y) \mid 2x + y > 2\}$
4.  $\{(x, y) \mid 3x - y > 2\} \cap \{(x, y) \mid x + 3y > {}^-6\}$
5.  $\{(x, y) \mid x - y > 5\} \cap \{(x, y) \mid x - y > 2\}$
6.  $\{(x, y) \mid x + 2y < {}^-1\} \cap \{(x, y) \mid 3x - 2y > 2\}$
7.  $\{(x, y) \mid 2x + y < 1\} \cap \{(x, y) \mid 2x + y > 3\}$
8.  $\{(x, y) \mid x - 4y > 2\} \cap \{(x, y) \mid x + 2y < 6\}$
9.  $\{(x, y) \mid x + 2y > 4\} \cap \{(x, y) \mid x + 2y < {}^-2\}$
10. $\{(x, y) \mid 2x + 3y < 6\} \cap \{(x, y) \mid 2x + 3y < {}^-3\}$

# 6 ALGEBRAIC POLYNOMIALS, FACTORING, AND FRACTIONS

## 6.1 Polynomials

Algebraic expressions such as $5x^2 - 3x + 2$ and $2x^2y - 3xy + 7xy^4 - 2$ are called *polynomials*. In general, any algebraic expression which contains at least one indicated addition and has no variables in any denominator will be called a polynomial.

An expression, algebraic or numerical, which has multiplication as its last operation, such as $4x^2y$ or $5x(x - 3)$ is called a monomial. $^-2xy^3z^7$ and $7x$ are two other examples of monomials. The distinguishing feature of a monomial is the fact that its last operation must be multiplication, however, this feature amounts to the restriction that a monomial cannot contain indicated addition as its last operation.

$4x^2 + 3y$ is not a monomial because it has an indicated addition and addition is the last operation to be performed. $4x^2$, by itself, is a monomial; $3y$, by itself, is also a monomial; the indicated sum, $4x^2 + 3y$, is not a monomial.

We say that $4x^2 + 3y$ has two *terms*. By this we mean that it consists of two monomials separated by an addition symbol. $5x + 3x^2y + 4y^3$ is an algebraic expression with three *terms* because it is the indicated sum of three monomials: $5x$, $3x^2y$, and $4y^3$.

The minus sign in $2x - 7y$ must be recognized as meaning the same as $2x + {}^-7y$. Hence, the minus sign is an indicated addition symbol and $2x - 7y$ has two terms, the monomials $2x$ and ${}^-7y$.

EXERCISE 6.1.1

State the number of terms in each of the expressions.

1. $9x + 3y^2$
2. $2x + 5y + 4x^2$
3. $2x + 3$
4. $9y - 7x^2 + 3z$
5. $4x^2y^5z$

6. $4x^2 - 3x + 5y - 2z^3$
7. $9 - 3x^2y$
8. ${}^-6xyzw^2$
9. $8x^2 - 3 + 5x^2$
10. $2xy + 5x - 3x - 4xy$

An expression with only one term is called a monomial. Any expression having more than one term is called a polynomial.

$3x^2 - 5x$ is a polynomial with two terms. Whenever a polynomial has exactly two terms it may be called a *binomial*.

$x^2 + 3x - 7$ is a polynomial with three terms. Whenever a polynomial has exactly three terms it may be called a *trinomial*.

"Monomial," "binomial," and "trinomial" are special words which refer to a specific number of terms in an algebraic expression. "Polynomial" refers to an algebraic expression with two or more terms.

EXERCISE 6.1.2

True or false:

1. A binomial must have exactly two terms.
2. Every trinomial is also a polynomial.
3. Some polynomials are monomials.
4. Some polynomials are binomials.
5. Every polynomial is a trinomial.
6. No binomial is a trinomial.
7. No trinomial is a polynomial.
8. ${}^-5x^2y^3z$ is a monomial.
9. $2x - 3$ is a binomial.
10. $x^2 + 3x - 2$ is a trinomial.

## 6.2  Simplifying Polynomials

$2x - 13x$ is a binomial in which both terms have exactly the same variable, $x$. The Distributive Law of Multiplication over Addition is used to simplify $2x - 13x$ as follows:

$$2x - 13x$$

$$(2 - 13)x$$

$$^-11x$$

Hence the binomial $2x - 13x$ can be simplified to the monomial $^-11x$.

$3x + 2y$ is another binomial, but in this case the terms have different variables. This difference in variables makes it impossible to simplify $3x + 2y$ because the Distributive Law of Multiplication over Addition does not apply to such expressions. The binomial $3x + 2y$ can not be simplified to a monomial.

The binomial $5x^2 + 3x$ can not be simplified to a monomial even though each of the terms has the variable $x$. This is because the variable does not have the same exponent in each term. To attempt to simplify $5x^2 + 3x$ using the Distributive Law of Multiplication over Addition we are led to the following result:

$$5x^2 + 3x$$

$$(5x + 3)x$$

Since $5x + 3$ can not be simplified, then $(5x + 3)x$ also can not be simplified. Therefore, $5x^2 + 3x$ can not be simplified to a monomial.

$4x^3y - 7x^3y$ is a binomial that can be simplified to a monomial. This is because the variables and the exponents for those variables are exactly the same. $4x^3y - 7x^3y$ is simplified by the Distributive Law.

$$4x^3y - 7x^3y$$

$$(4 - 7)x^3y$$

$$^-3x^3y$$

Terms which have exactly the same variable and exactly the same exponent for each of the individual variables are called *like* terms.

$2x^5y^3z^7$ and $^-15x^5y^3z^7$ are *like* terms because they both involve the variables $x$, $y$, $z$, and the exponents for the variables in both terms are 5 for the $x$, 3 for the $y$, and 7 for the $z$.

$9x^3y^4z^2$ and $4x^3y^5z^2$ are not *like* terms. Although they involve the same variables, the fact that the $y$'s have different exponents determines $9x^3y^4z^2$ and $4x^3y^5z^2$ are not *like* terms.

Whenever a polynomial has like terms then it can be simplified as shown in the following example. Note that the like terms are first grouped together. This reordering and regrouping is justified by the Commutative and Associative Laws of Addition.

$$2x^2y - 6xy^5 - 5x^2y + xy^5$$

$$(2x^2y - 5x^2y) + (^-6xy^5 + xy^5)$$

$$(2 - 5)x^2y + (^-6 + 1)xy^5$$

$$^-3x^2y - 5xy^5$$

EXERCISE 6.2.1

Simplify wherever possible:

1.  $9x - 5x$
2.  $3x^2 + 7x^2$
3.  $6x - 1$
4.  $4x^2y + 3x^2y$
5.  $^-5xy + xy$
6.  $3x^2 + 2x$
7.  $9x^3y^5 - 6x^3y^5$
8.  $4x^2 - 3x^2 + 9x^2$
9.  $6xy^3z^4 + xy^3z^4 - 4xy^3z^4$
10. $3x^2y^5 - 2x^5y^2$
11. $4x^2 - 3y + 2x^2 + y$
12. $^-3xy + 7x - 2xy - 3x$
13. $x^2 + 7x + 2x + 14$
14. $3xy^2 - 2xy^2 + 4x - xy^2$
15. $5x^2 - 3x + 4 + 2x$
16. $^-3 + 4x - 3x^2 + 2$
17. $9xy^5z^2 - 3xy^2z + 2xy^5z^2 - 3xyz^2$
18. $4x^3 - 3x^2 + 4x^2 - 2x + 7x^2$
19. $6x^2y - 2xy^2 - 3xy + 2xy^2$
20. $5x^3 - 2x + 4x^2 - 3x + 5x^3 - 2 + 6$

## 6.3  Multiplying Monomials

The multiplication of monomials is completely explainable in terms of the Commutative and Associative Laws of Multiplication. The Commutative Law of Multiplication allows us to change the order of any factors. The Associative Law of Multiplication allows us to group our factors in whatever way we please.

To multiply $3x^2y$ and $^-5xy^5$ the problem is completed in the following steps:

$$3x^2y \cdot {}^-5xy^5$$

$$(3 \cdot {}^-5) \cdot (x^2 \cdot x) \cdot (y \cdot y^5)$$

$$^-15x^3y^6$$

Notice that the factors of $3x^2y$ and $^-5xy^5$ are reordered and regrouped so that there is one numerical product, $(3 \cdot {}^-5)$, and an indicated product for each variable, $(x^2 \cdot x)$ and $(y \cdot y^5)$.

Remember how the multiplication of variables with exponents is accomplished. $x^2$ means $x \cdot x$ because the exponent 2 means that there are two factors of $x$. $x^2 \cdot x$ means $(x \cdot x) \cdot x$ and is simplified to $x^3$ because there is a total of three factors of $x$. An exponent shows the count of the number of factors. Because of this use of exponents, problems such as $y \cdot y^5$ and $z^3 \cdot z^6$ are simplified to $y^6$ and $z^9$. The exponents of the examples $y \cdot y^5 = y^6$ and $z^3 \cdot z^6 = z^9$ were added. This is because the exponents show the total number of factors involved in the two expressions.

As another example of the multiplication of monomials consider the following:

$$(x^2y^5z^2) \cdot ({}^-x^3y^2z^9)$$

$$(1x^2y^5z^2) \cdot ({}^-1x^3y^2z^9)$$

$$(1 \cdot {}^-1) \cdot (x^2 \cdot x^3) \cdot (y^5 \cdot y^2) \cdot (z^2 \cdot z^9)$$

$$({}^-1) \cdot (x^5) \cdot (y^7) \cdot (z^{11})$$

$$^-x^5y^7z^{11}$$

The preceding example shows the use of 1 and $^-1$ as factors and also should be studied for its illustration of multiplication using exponents.

EXERCISE 6.3.1

Multiply:

| | | | |
|---|---|---|---|
| 1. | $x^4 \cdot x^3$ | 11. | $4x^2y^2 \cdot {}^-xy^4$ |
| 2. | $z^4 \cdot z$ | 12. | $^-2x^3y \cdot 5x^5y^2$ |
| 3. | $y^4 \cdot {}^-y^2$ | 13. | $3xy^4 \cdot {}^-5x^2y^5$ |
| 4. | $^-z^5 \cdot {}^-z^3$ | 14. | $^-4xy^2 \cdot {}^-x^3y^3$ |
| 5. | $2x^2 \cdot 3x^5$ | 15. | $^-7xy \cdot 4xy$ |
| 6. | $3x^4 \cdot 5x^2$ | 16. | $9x^2y^3z \cdot 2xy^2z^6$ |
| 7. | $4x \cdot {}^-2x^3$ | 17. | $^-3xyz^4 \cdot {}^-x^2yz^5$ |
| 8. | $4x^5 \cdot 6x^3$ | 18. | $x^{15} \cdot x^9$ |
| 9. | $^-7y^3 \cdot {}^-3y^2$ | 19. | $y^{41} \cdot y^{67}$ |
| 10. | $^-5x^3y^2 \cdot {}^-x^2y$ | 20. | $z^{97} \cdot z^{205}$ |

### 6.4  Removing Parentheses

The parentheses of expressions such as $6(x - 3y)$ are removed by applying the Distributive Law of Multiplication over Addition.

The first step in removing the parentheses from $6(x - 3y)$ is to distribute the 6 over the two terms of the binomial $x - 3y$.

$$6(x - 3y)$$

$$6 \cdot x + 6 \cdot {}^-3y$$

$$6x + {}^-18y$$

$$6x - 18y$$

Notice that the second line of the example $(6 \cdot x + 6 \cdot {}^-3y)$ involves the multiplication of monomials. Hence the removal of the parentheses from $6(x - 3y)$ can be viewed as the multiplication of each term of the polynomial, $x - 3y$, by the monomial, 6.

The parentheses of $^-2x^2y(x - 7y^2z)$ are removed by multiplying each term of the polynomial, $x - 7y^2z$, by the monomial, $^-2x^2y$.

$$^-2x^2y(x - 7y^2z)$$

$$^-2x^2y \cdot x + {}^-2x^2y \cdot {}^-7y^2z$$

$$^-2x^3y + 14x^2y^3z$$

The removal of the parentheses from $(2x^2 - 7x - 3) \cdot 5x$ can be used as an example to illustrate two facets of this skill. First, the monomial may be situated at either the right or left end of the parentheses without altering the procedure. Secondly, the polynomial may have more than two terms.

$$(2x^2 - 7x - 3) \cdot 5x$$

$$2x^2 \cdot 5x + {}^-7x \cdot 5x + {}^-3 \cdot 5x$$

$$10x^3 + {}^-35x^2 + {}^-15x$$

$$10x^3 - 35x^2 - 15x$$

In every case where a polynomial is to be multiplied by a monomial, each term of the polynomial is multiplied by the monomial.

EXERCISE 6.4.1

Remove the parentheses:

1.  $2(x - 6)$
2.  $^-5(2x - 3)$

3. $^-(x-1)$   [Hint: $^-(x-1)$ is equivalent to $^-1(x-1)$.]
4. $4(x^2-3x+2)$
5. $7(x+8)$
6. $^-2(3x-7)$
7. $6(2x+8)$
8. $^-3(2x+5)$
9. $^-(4x-7)$
10. $3x(4x-1)$
11. $^-5x(3x-8)$
12. $^-3x(3x-7)$

13. $6xy^2(3x-4y^5)$
14. $2x^3y(3x^4y^3-5x^5)$
15. $(x+7)\cdot{}^-2$
16. $(2x-3)\cdot{}^-4x$
17. $(4-5x)\cdot{}^-x$
18. $4x^2(3x^2-x+3)$
19. $(x^2-3x+2)\cdot{}^-4$
20. $^-(7x^2-6x+4)$

## 6.5   Multiplying Polynomials

The removal of the parentheses from $^-5x(3x-2)$ is accomplished by multiplying each term of $3x-2$ by $^-5x$.

The same procedure is used to remove the parentheses from $(x+4)(y-3)$, but in this case each term of $y-3$ must be multiplied by $(x+4)$.

$$(x+4)(y-3)$$

$$(x+4)\cdot y + (x+4)\cdot{}^-3$$

$$x\cdot y + 4\cdot y + x\cdot{}^-3 + 4\cdot{}^-3$$

$$xy + 4y - 3x - 12$$

Study the preceding problem carefully. Notice the manner in which $(x+4)$ is distributed over each term of $(y-3)$. This same procedure is used in removing the parentheses in the following example.

$$(x-2y)(z+6w-4k)$$

$$(x-2y)\cdot z + (x-2y)\cdot 6w + (x-2y)\cdot{}^-4k$$

$$x\cdot z + {}^-2y\cdot z + x\cdot 6w + {}^-2y\cdot 6w + x\cdot{}^-4k + {}^-2y\cdot{}^-4k$$

$$xz - 2yz + 6wz - 12wy - 4kx + 8ku$$

EXERCISE 6.5.1

Multiply:

1. $(y+5)(x+2)$
2. $(x+6)(y-3)$
3. $(x-4)(y+9)$
4. $(x-2)(y-5)$
5. $(x+6)(y+3)$
6. $(y-2)(x-2)$

7. $(x^2+3)(x+1)$
8. $(x-6)(x^2-2)$
9. $(3x-2)(y+4)$
10. $(2x-5)(3y+2)$
11. $(9x-1)(2y+3)$
12. $(3x+4)(5y-6)$

13.  $(2x - 3)(4y + 7)$          17.  $(5x - 4)(5y + 2)$
14.  $(5x + 2)(2y - 5)$          18.  $(x - 2)(3y - 2z + w)$
15.  $(7x - 1)(y + 3)$           19.  $(2x + 3)(y + 4z - 3zw)$
16.  $(3x + 8)(2y - 6)$          20.  $(3x - 5)(2z - y + 1)$

## 6.6  Multiplying Polynomials and Simplifying Results

In the preceding exercise whenever the polynomials were multiplied no like terms were obtained and, consequently, no simplification of the results was possible. Oftentimes the polynomials we wish to multiply will produce like terms and will require simplification. An example of this type of problem is shown below.

$$(2x - 3y^2)(5x + y^2)$$

$$(2x - 3y^2) \cdot 5x + (2x - 3y^2) \cdot y^2$$

$$10x^2 - 15xy^2 + 2xy^2 - 3y^4$$

$$10x^2 - 13xy^2 - 3y^4$$

In the third step of the preceding problem two like terms, $^-15xy^2$ and $2xy^2$, were obtained. These terms should be simplified to $^-13xy^2$ as shown in the last line of the example.

As another illustration of the type of multiplication which produces like terms and should have the results simplified, study the following example.

$$(x - 3y)(x^2 - 2xy + 5y^2)$$

$$(x - 3y) \cdot x^2 + (x - 3y) \cdot {}^-2xy + (x - 3y) \cdot 5y^2$$

$$x^3 - 3x^2y - 2x^2y + 6xy^2 + 5xy^2 - 15y^3$$

$$x^3 + (^-3x^2y - 2x^2y) + (6xy^2 + 5xy^2) - 15y^3$$

$$x^3 - 5x^2y + 11xy^2 - 15y^3$$

To assist you in following the preceding example, the fourth line shows the grouping, by parentheses, of like terms. Remember, like terms must have exactly the same variables and each variable must have the same exponent as its counterpart in the other term.

EXERCISE 6.6.1

Multiply and simplify:

1.  $(4x - 3)(2x + 3)$          3.  $(4x - 5)(x - 3)$
2.  $(3x + 1)(5x + 2)$          4.  $(x + 5)(3x - 2)$

| | |
|---|---|
| 5.  $(x + 4)(x + 3)$ | 13.  $(x + 5)(x^2 + 3x - 2)$ |
| 6.  $(x + 5)(x - 8)$ | 14.  $(x - 4)(x^2 - 3x + 4)$ |
| 7.  $(x - 7)(x + 4)$ | 15.  $(2x + 3y)(3x - 7y)$ |
| 8.  $(x + 4)(x - 5)$ | 16.  $(x^2 + 10y)(x^2 - 10y)$ |
| 9.  $(x - 10)(x + 3)$ | 17.  $(2x - y)(x + 5y)$ |
| 10.  $(x + 2)(x - 2)$ | 18.  $(2x - 7y)(2x + 7y)$ |
| 11.  $(x + 5)(x^2 + 2x + 3)$ | 19.  $(x + 6y)(x^2 - 3xy - y^2)$ |
| 12.  $(2x + 3)(x^2 + 5x - 7)$ | 20.  $(x - 2y)(x^2 + 4xy - 3y^2)$ |

## 6.7   Multiplying Binomials (A Short Method)

The multiplication of binomials arises so frequently throughout the remainder of your work in Algebra that it behooves us to suggest a useful short-cut that may be used. In learning this short-cut method, however, you must remember that it only applies to a particular type of problem. The methods used in the last two exercises may be used in the multiplication of any two polynomials; the method suggested in this section only applies to the multiplication of special types of binomials which occur frequently in your future work.

Consider the following multiplication:

$$(2x - 7)(x + 2)$$

$$(2x - 7) \cdot x + (2x - 7) \cdot 2$$

$$2x^2 - 7x + 4x - 14$$

$$2x^2 - 3x - 14$$

In the answer $2x^2$ is the first term. What part of the multiplication produced the $2x^2$? It is the product of the first terms of the two binomials.

$$(2x - 7)(x + 2) \qquad 2x \cdot x = 2x^2$$

In the answer $2x^2 - 3x - 14$, what part of the problem $(2x - 7)(x + 2)$ produced the ⁻14? It is the product of the last terms of the two binomials.

$$(2x - 7)(x + 2) \qquad {}^-7 \cdot 2 = {}^-14$$

In the answer $2x^2 - 3x - 14$, what parts of the problem $(2x - 7)(x + 2)$

produced the $^-3x$? It is the sum of the products of the "inner" and "outer" terms.

$$(2x - 7)(x + 2) \qquad 4x + {}^-7x = {}^-3x$$

$$(2x - 7)(x + 2) \qquad 2x^2 - 3x - 14$$

Study the following problem carefully.

$$(4x + 1)(3x - 5)$$
$$(4x + 1)\cdot 3x + (4x + 1)\cdot {}^-5$$
$$12x^2 + 3x - 20x - 5$$
$$12x^2 - 17x - 5$$

$12x^2$ is the result of multiplying the first two terms of the binomials.

$$(4x + 1)(3x - 5) \qquad 4x \cdot 3x = 12x^2$$

$^-17x$ is the result of adding the products of the "inner" and "outer" terms of $(4x + 1)(3x - 5)$.

$$(4x + 1)(3x - 5) \qquad {}^-20x + 3x = {}^-17x$$

$^-5$ is the result of multiplying the last two terms of $(4x + 1)(3x - 5)$.

$$(4x + 1)(3x - 5) \qquad 1\cdot {}^-5 = {}^-5$$

The product of $(3x - 2)(x + 6)$ is a trinomial whose terms are acquired as shown in the following figure.

$$(3x - 2)(x + 6) \qquad 3x^2 + 18x - 2x - 12$$
$$3x^2 + 16x - 12$$

Whenever two binomials are to be multiplied and the first terms are like terms and the second terms are like terms then:

(1)   The first term of the product is the product of the first terms of the binomials.

(2)   The middle term of the product is the sum of the products of the "inner" and "outer" terms.

(3)   The last term of the product is the product of the second terms of the binomials.

With a little practice, the multiplication of binomials can be completely accomplished without writing down any of the intermediate steps. In the following exercise try doing the complete multiplication in your head; until you are certain you are getting correct answers you should check by multiplying in the way used in previous exercises.

EXERCISE 6.7.1

Multiply:

| | |
|---|---|
| 1.  $(x + 9)(x - 3)$ | 14.  $(x + 6)(x - 6)$ |
| 2.  $(x + 7)(x - 4)$ | 15.  $(x - 8)(x + 2)$ |
| 3.  $(x - 5)(x + 6)$ | 16.  $(x + 7)(x + 3)$ |
| 4.  $(2x + 3)(x + 4)$ | 17.  $(2x - 5)(2x + 5)$ |
| 5.  $(x + 2)(3x + 4)$ | 18.  $(3x + 1)(2x + 3)$ |
| 6.  $(4x - 3)(x + 4)$ | 19.  $(x - 7)^2$ or $(x - 7)(x - 7)$ |
| 7.  $(x + 6)(x + 2)$ | 20.  $(x + 7)(x - 1)$ |
| 8.  $(x + 5)(x + 8)$ | 21.  $(x + 8)^2$ |
| 9.  $(x + 5)(x + 4)$ | 22.  $(x + 4)(x - 4)$ |
| 10.  $(3x - 10)(x + 1)$ | 23.  $(2x + 7)^2$ |
| 11.  $(x - 3)(x + 3)$ | 24.  $(5x - 1)(5x + 1)$ |
| 12.  $(2x - 1)(x + 5)$ | 25.  $(x - 6)(x - 5)$ |
| 13.  $(4x - 7)(2x - 1)$ | |

## 6.8  Adding Algebraic Fractions

The addition of $\dfrac{5}{8} + \dfrac{3}{7}$ is accomplished by

$$\frac{5}{8} + \frac{3}{7} = \frac{5 \cdot 7 + 8 \cdot 3}{8 \cdot 7}$$

$$= \frac{35 + 24}{56}$$

$$= \frac{59}{56}$$

The same procedure is used to add $\dfrac{4x}{3y} + \dfrac{2}{5x}$ .

$$\frac{4x}{3y} + \frac{2}{5x} = \frac{4x \cdot 5x + 3y \cdot 2}{3y \cdot 5x}$$

$$= \frac{20x^2 + 6y}{15xy}$$

Whenever the numerator or denominator is a polynomial the addition of algebraic fractions may involve the multiplication of binomials.

$$\frac{x-2}{x+5} + \frac{x+1}{x+2} = \frac{(x-2)(x+2) + (x+5)(x+1)}{(x+5)(x+2)}$$

$$= \frac{x^2 - 4 + x^2 + 6x + 5}{x^2 + 7x + 10}$$

$$= \frac{2x^2 + 6x + 1}{x^2 + 7x + 10}$$

The last example to be offered here is to illustrate the use of the minus sign as it applies to the addition of algebraic fractions.

Recall that $\dfrac{3}{4} - \dfrac{7}{8}$ means $\dfrac{3}{4} + \dfrac{^-7}{8}$ . We continue to use the minus sign as an indication that we are to add the opposite of its following term. Study the following example.

$$\frac{x-3}{x+3} - \frac{x+5}{x-2} = \frac{x-3}{x+3} + \frac{^-(x+5)}{x-2}$$

$$= \frac{(x-3)(x-2) + {}^-1(x+3)(x+5)}{(x+3)(x-2)}$$

$$= \frac{x^2 - 5x + 6 + {}^-1(x^2 + 8x + 15)}{x^2 + x - 6}$$

$$= \frac{x^2 - 5x + 6 - x^2 - 8x - 15}{x^2 + x - 6}$$

$$= \frac{^-13x - 9}{x^2 + x - 6}$$

EXERCISE 6.8.1

Complete the indicated addition:

1. $\dfrac{3}{4} + \dfrac{1}{5}$

2. $\dfrac{2}{3} - \dfrac{1}{4}$

3. $\dfrac{3}{4} + \dfrac{1}{7}$

4. $\dfrac{5}{9} - \dfrac{2}{5}$

5. $\dfrac{3x}{4} + \dfrac{1}{y}$

6. $\dfrac{3}{x} + \dfrac{4}{y}$

7. $\dfrac{7}{x} - \dfrac{2}{y}$

8. $\dfrac{x-4}{5} + \dfrac{2}{7}$

9. $\dfrac{x+3}{x-2} + \dfrac{1}{3}$

10. $\dfrac{x-4}{2} - \dfrac{2}{5}$

11. $\dfrac{x+2}{3} + \dfrac{1}{x-3}$

12. $\dfrac{x-7}{5} + \dfrac{3}{x-2}$

13. $\dfrac{x-9}{x+1} + \dfrac{x+2}{x-1}$

14. $\dfrac{x+4}{x-1} - \dfrac{x-3}{x+6}$

15. $\dfrac{x-9}{x-2} + \dfrac{x+4}{x+3}$

16. $\dfrac{x-7}{x+2} + \dfrac{x-5}{x-6}$

17. $\dfrac{2x+1}{x-3} + \dfrac{x-1}{x+3}$

18. $\dfrac{3x-4}{x+1} - \dfrac{x+2}{x+4}$

19. $\dfrac{x-3}{x+4} + \dfrac{2x-5}{x-3}$

20. $\dfrac{2x+5}{x-3} + \dfrac{3x+1}{x-4}$

## 6.9   Subtraction of Polynomials

In Algebra subtraction simply means to add the opposite. Problems such as $4 - 7$ and $3 - {}^-8$ are written as $4 + {}^-7$ and $3 + 8$. In each case the opposite of the second number was added to the first.

$(3x^2 + 5x - 6) - (x^2 - 2x + 1)$ can be viewed as a subtraction problem, but our use of the minus sign still means that we can add the opposite of $(x^2 - 2x + 1)$ to $(3x^2 + 5x - 6)$.

What is the opposite of $(x^2 - 2x + 1)$? Since the opposite of $(x^2 - 2x + 1)$ can be written as $^-1(x^2 - 2x + 1)$, we can multiply each term by $^-1$ to obtain $^-x^2 + 2x - 1$.

Notice that the opposite of any polynomial can be obtained by changing the sign of each term.

$$3x^2 - 5x + 2 \text{ is the opposite of } ^-3x^2 + 5x - 2$$

$$^-5x - 4 \text{ is the opposite of } 5x + 4$$

$$^-2x + 3 \text{ is the opposite of } 2x - 3$$

Sometimes we are given directions to subtract one polynomial from another. Whenever these directions are given then the opposite of the second polynomial is to be added to the first polynomial.

The following is an example of a subtraction problem with polynomials.

$$\text{Subtract:} \quad \begin{array}{r} 5x^2 - 3x - 2 \\ 2x^2 - 4x + 9 \\ \hline \end{array}$$

This problem is properly done by first finding the opposite of the second polynomial and then adding. Hence the problem is completed as

$$\begin{array}{r} 5x^2 - 3x - 2 \\ ^-2x^2 + 4x - 9 \\ \hline 3x^2 + x - 11 \end{array}$$

EXERCISE 6.9.1

1. Find the opposite of each polynomial.

   (a) $2x + 6$        (e) $4x - 6$

   (b) $3x - 9$        (f) $x^2 - 9x + 2$

   (c) $^-3x + 4$        (g) $^-3x^2 + 2x + 1$

   (d) $5x + 7$        (h) $4x^2 - 3x - 5$

2. Do each of the following as a subtraction problem.

   (a) $\begin{array}{r} 5x - 3 \\ 3x + 4 \\ \hline \end{array}$        (c) $\begin{array}{r} 3x - 7 \\ 4x + 2 \\ \hline \end{array}$

   (b) $\begin{array}{r} 4x + 9 \\ 2x - 5 \\ \hline \end{array}$        (d) $\begin{array}{r} x - 3 \\ 3x - 5 \\ \hline \end{array}$

(e)   $x^2 + 4x$
      $2x^2 - 5x$

(h)   $3x + 5$
      $3x - 9$

(f)   $8x^2 - 5x$
      $8x^2 - 5x$

(i)   $12x^2 - 8x$
      $12x^2 + 8x$

(g)   $x^2 - 5x$
      $x^2 + 2x$

(j)   $3x - 8$
      $5x - 2$

## 6.10   Long Division of Polynomials

The long division of polynomials is accomplished in a series of steps very similar to the numerical problem shown below.

$$
\begin{array}{r}
61 \\
34\overline{)2085} \\
\underline{204} \\
45 \\
\underline{34} \\
11
\end{array}
$$

34 is the divisor
2085 is the dividend
11 is the remainder

We can best explain the skill to be developed in this section by reviewing the step-by-step procedure used on the preceding problem and comparing it to a problem involving the long division of polynomials.

Compare the following two problems.

$$34\overline{)2085} \qquad x - 3\overline{)x^2 + 7x - 11}$$

(1)   The first step in the numerical problem is to estimate the number of 34's in 208. This is often done by estimating the number of 3's in 20. Similarly, to begin the polynomial problem we divided

$x^2$ by $x$. $\dfrac{x^2}{x} = x$. Hence the first steps for the two problems are

completed as shown below.

$$
\begin{array}{r}
6 \\
34\overline{)2085}
\end{array}
\qquad
\begin{array}{r}
x \\
x - 3\overline{)x^2 + 7x - 11}
\end{array}
$$

(2)   Although we used $20 \div 3$ to arrive at the 6 in the answer we now multiply 6 by 34 and place this product below 208. Similarly,

we now multiply $x$ by $(x - 3)$. The second step is completed as shown below.

$$
\begin{array}{r}
6 \\
\hline
34)\overline{2085} \\
204
\end{array}
\qquad
\begin{array}{r}
x \\
\hline
x - 3)\overline{x^2 + 7x - 11} \\
x^2 - 3x
\end{array}
$$

(3)    In the numerical problem we now subtract 204 from 208 and "bring down" the 5. In the polynomial problem we also must subtract. Remembering that algebraic subtraction is addition of the opposite, we arrive at the following results.

$$
\begin{array}{r}
6 \\
\hline
34)\overline{2085} \\
204 \\
\hline
45
\end{array}
\qquad
\begin{array}{r}
x \\
\hline
x - 3)\overline{x^2 + 7x - 11} \\
x^2 - 3x \\
\hline
10x - 11
\end{array}
$$

(4)    For the numerical problem we now divide 34 into 45. For the polynomial problem we now divide $x - 3$ into $10x - 11$ by dividing $10x$ by $x$.

$$\frac{10x}{x} = 10$$

$$
\begin{array}{r}
61 \\
\hline
34)\overline{2085} \\
204 \\
\hline
45
\end{array}
\qquad
\begin{array}{r}
x + 10 \\
\hline
x - 3)\overline{x^2 + 7x - 11} \\
x^2 - 3x \\
\hline
10x - 11
\end{array}
$$

(5)    Multiplying the divisors by 1 and 10 respectively we obtain

$$
\begin{array}{r}
61 \\
\hline
34)\overline{2085} \\
204 \\
\hline
45 \\
34
\end{array}
\qquad
\begin{array}{r}
x + 10 \\
\hline
x - 3)\overline{x^2 + 7x - 11} \\
x^2 - 3x \\
\hline
10x - 11 \\
10x - 30
\end{array}
$$

(6)    Subtracting, we arrive at our final answers which show the remainders of 11 and 19 respectively.

$$
\begin{array}{r}
61 \\
\hline
34)\overline{2085} \\
204 \\
\hline
45 \\
34 \\
\hline
11
\end{array}
\qquad
\begin{array}{r}
x + 10 \\
\hline
x - 3)\overline{x^2 + 7x - 11} \\
x^2 - 3x \\
\hline
10x - 11 \\
10x - 30 \\
\hline
19
\end{array}
$$

(7)   The problems are checked by multiplying the answer by the divisor and adding the remainder to the product. The result must be the dividend.

$$\begin{array}{r} 61 \\ \times\ 34 \\ \hline 244 \\ 183 \\ \hline 2074 \\ +\ \ 11 \\ \hline 2085 \end{array}$$

$(x + 10)(x - 3) = x^2 + 7x - 30$

$(x^2 + 7x - 30) + 19 = x^2 + 7x - 11$

Here is a complete example of a long division problem with polynomials

$$x - 3\overline{)x^2 - 7x + 14} \quad \begin{array}{r} x - 4 \end{array}$$

$$\begin{array}{r} \underline{x^2 - 3x} \\ {}^-4x + 14 \\ \underline{{}^-4x + 12} \\ 2 \end{array}$$

Check:

$(x - 3)(x - 4) + 2$

$(x^2 - 7x + 12) + 2$

$x^2 - 7x + 14$

The long division of $x^3 - 3x + 4$ by $x + 2$ is shown below. Notice that an extra term, $0x^2$, has been included in the dividend. Study the problem and its check carefully to insure that you completely understand the procedure involved in the long division of polynomials.

$$x + 2\overline{)x^3 + 0x^2 - 3x + 4} \quad \begin{array}{r} x^2 - 2x + 1 \end{array}$$

$$\begin{array}{r} \underline{x^3 + 2x^2} \\ {}^-2x^2 - 3x \\ \underline{{}^-2x^2 - 4x} \\ x + 4 \\ \underline{x + 2} \\ 2 \end{array}$$

Check:

$(x + 2)(x^2 - 2x + 1) = x^3 - 3x + 2$

$(x^3 - 3x + 2) + 2 = x^3 - 3x + 4$

EXERCISE 6.10.1

Divide and check:

1.  $x - 5\overline{)x^2 + 2x - 3}$

2.  $x + 3\overline{)x^2 - 5x + 1}$

3.  $x - 2\overline{)x^2 - 6x + 8}$

4.  $x + 1\overline{)x^3 + 2x^2 - x + 3}$

5.  $x - 4\overline{)x^3 - 3x^2 - 5x + 2}$

6.  $x - 3\overline{)x^3 - 5x^2 - 5x - 3}$

7.  $x + 2\overline{)x^3 + 2x - 1}$

8.  $x + 7\overline{)x^2 + 5x - 14}$

9.  $x - 2\overline{)x^2 - 4}$

10.  $x - 6\overline{)x^2 - 7x + 6}$

11. $x + 3\overline{)x^2 + 7x + 14}$    16. $x - 8\overline{)x^2 - 64}$

12. $x - 6\overline{)x^2 - 7x + 6}$    17. $x - 1\overline{)x^2 + 5x - 7}$

13. $x + 2\overline{)x^2 + 9x - 3}$    18. $x - 3\overline{)x^2 - 10x + 3}$

14. $x - 4\overline{)x^3 - 3x^2 - 2x - 8}$    19. $x + 8\overline{)x^2 + 7x - 8}$

15. $x + 5\overline{)x^3 + 7x^2 + 7x - 15}$    20. $x - 9\overline{)x^2 + x + 1}$

## 6.11 Factors of a Monomial

The word "factor" refers to a number or algebraic expression that is part of multiplication.

When we say that 2 is a factor of 10 we simply mean that there is an integer that can be multiplied by 2 to give 10. We say 3 is not a factor of 10 because the use of the word "factor" is limited here to integers. Were rational numbers allowed as factors then every nonzero rational number would be a factor of 10. Why?

2 is a factor of 10 because $2 \cdot 5 = 10$. This same multiplication indicates that 5 is also a factor of 10. But 2 and 5 are not the only factors of 10 because there are three other pairs of integers that have a product of 10. $1 \cdot 10 = 10$, $^-1 \cdot {}^-10 = 10$, and $^-2 \cdot {}^-5 = 10$. Therefore, the set of all factors of 10 is $\{10, 5, 2, 1, {}^-1, {}^-2, {}^-5, {}^-10\}$.

Every integer will have itself, its opposite, 1, and $^-1$ as a factor. Whenever these are the only factors of an integer then it is called a *prime* number. Whenever an integer has other factors than itself, its opposite, 1, and $^-1$ then it is called a *composite* number.

The set of all factors of 24 is

$$\{24, 12, 8, 6, 4, 3, 2, 1, {}^-1, {}^-2, {}^-3, {}^-4, {}^-6, {}^-8, {}^-12, {}^-24\}$$

Since the set includes numbers other than 24, $^-24$, 1, and $^-1$ we say that 24 is a composite number.

The set of all factors of 13 is

$$\{13, 1, {}^-1, {}^-13\}$$

Since this set only contains the number, its opposite, 1, and $^-1$ we say that 13 is a prime number.

The set of all variable factors of $x^2$ is $\{x, x^2\}$.

The set of all variable factors of $y^3$ is $\{y, y^2, y^3\}$.

The set of all variable factors of $x^2y^3$ must include all the factors of $x^2$ and $y^3$ in every possible combination that can be made of such factors. Hence the set of all variable factors of $x^2y^3$ is $\{x, x^2, y, y^2, y^3, xy, x^2y, xy^2, x^2y^2, xy^3, x^2y^3\}$.

$5x^3y^2$ is a monomial expression involving the integer 5 and the variables $x$ and $y$. The set of all factors of 5 is $\{5, 1, {}^-1, {}^-5\}$. The set of all variable factors of $5x^3y^2$ is $\{x^3, x^2, x, y^2, y, x^3y, x^3y^2, x^2y^2, x^2y, xy^2, xy\}$.

EXERCISE 6.11.1

For each monomial find (1) the set of all numerical factors, (2) the set of all variable factors, and (3) whether the integer is prime or composite.

1. $8xy$
2. $^-9x^2$
3. $7xy^2$
4. $12x^2y^2$
5. $^-11x^3$

6. $20x$
7. $^-15xy^3$
8. $17x^4$
9. $27x^3y$
10. $^-30x^2y^4$

## 6.12   Factoring Polynomials with a Common Factor

In previous sections of this text problems such as $2x(x - 3y)$ have been presented with directions to remove the parentheses by multiplying each term of $x - 3y$ by $2x$.

$$2x(x - 3y) = 2x^2 - 6xy$$

This means that $2x^2 - 6xy$ is the product of multiplication and $2x$ and $(x - 3y)$ are therefore factors of $2x^2 - 6xy$. In this section you will learn how to find factors of polynomials such as $5x - 10$ and $2x^3 - 8x^2 + 12x$.

To factor $5x - 10$ the first question to be asked is "What factors, other than 1 and $^-1$, do $5x$ and $^-10$ have in common?" The answer to this question is 5 and $^-5$. Using 5 as a common factor for $5x$ and $^-10$ the following steps can be used:

$$5x - 10 = 5 \cdot x - 5 \cdot 2$$
$$= 5(x - 2)$$

Hence $5x - 10$ has factors of 5 and $(x - 2)$. This can be checked by multiplying $5(x - 2)$ to see whether it really produces $5x - 10$.

$$5(x - 2) = 5 \cdot x - 5 \cdot 2$$
$$= 5x - 10$$

Notice that the check given directly above is just the reverse of the steps shown earlier for factoring $5x - 10$.

To factor $2x^3 - 8x^2 + 12x$ the first question to ask is "What factors other than 1 and $^-1$ do $2x^3$, $^-8x^2$, and $12x$ have in common?" The common factors are 2, $^-2$, $x$, $^-x$, $2x$, and $^-2x$. Using $2x$ as the common factor, $2x^3 - 8x^2 + 12x$ is factored as

$$2x^3 - 8x^2 + 12x = 2x \cdot x^2 - 2x \cdot 4x + 2x \cdot 6$$
$$= 2x(x^2 - 4x + 6)$$

Hence $2x$ and $(x^2 - 4x + 6)$ are factors of $2x^3 - 8x^2 + 12x$. This is verified by multiplying the factors.

$$2x(x^2 - 4x + 6) = 2x \cdot x^2 - 2x \cdot 4x + 2x \cdot 6$$

$$= 2x^3 - 8x^2 + 12x$$

To factor $8x^2 - 12x$ it should be noticed that 2, 4, $2x$, and $4x$ are all common factors of $8x^2$ and $^-12x$. $4x$ should be used in the factoring because all the other common factors are also factors of $4x$. For this reason we say that $4x$ is the *largest common factor* for $8x^2$ and $^-12x$.

$$8x^2 - 12x = 4x \cdot 2x - 4x \cdot 3$$

$$= 4x(2x - 3)$$

The factors of $8x^2 - 12x$ are $4x$ and $(2x - 3)$.


EXERCISE 6.12.1

Factor by first finding the largest common factor for the terms of the polynomial. Check by multiplication.

| | | | |
|---|---|---|---|
| 1. | $6x + 12$ | 16. | $2x^2 - 16x + 4$ |
| 2. | $3x^2 - 9x$ | 17. | $3x^2 - 3x + 15$ |
| 3. | $4x + 6$ | 18. | $4x^2 - 2x + 10$ |
| 4. | $2x^2 + 6x - 10$ | 19. | $10x^2 + 15x + 25$ |
| 5. | $5x^2 - 10x$ | 20. | $6x^2 - 12x + 9$ |
| 6. | $14x + 35$ | 21. | $10x^3 - 3x^2 - 4x$ |
| 7. | $15x^2 - 40$ | 22. | $8x^2 - 16x - 8$ |
| 8. | $7x^2 - 2x$ | 23. | $6x^3 - 3x^2 + 30x$ |
| 9. | $9x + 24y$ | 24. | $4x^2 - 10x - 16$ |
| 10. | $8x - 18x^2$ | 25. | $9x^2 - 12x + 15$ |
| 11. | $12x + 14$ | 26. | $x^3 + 3x^2 - 11x$ |
| 12. | $19x - 21x^2$ | 27. | $xy - xz + xw$ |
| 13. | $7x - 63y^2$ | 28. | $5x^2 - 5z^3$ |
| 14. | $11x^2 + 33x$ | 29. | $x^2 + 3x$ |
| 15. | $9x^2 + 12$ | 30. | $7x - 7$ |

## 6.13    Other Situations Requiring Factoring by the Common Factoring Method

The most important method of factoring polynomials was practiced in the last exercise. Every other method of factoring to be studied in this text is only a more involved situation which depends upon an understanding of the common factoring method.

1 and $^-1$ are always common factors for any polynomial and occasionally it is important to use such factors. For example, $5x - 3$ can be factored as $1(5x - 3)$ using 1 and $(5x - 3)$ as factors. $^-6x + 7$ can be factored as $^-1(6x - 7)$ using $^-1$ and $(6x - 7)$ as factors.

$^-4x - 10$ could be factored as $2(^-2x - 5)$ or as $^-2(2x + 5)$. The preferred factorization is $^-2(2x + 5)$ because the first term of the binomial $(2x + 5)$ has a positive coefficient.

$^-8x + 4$ could be factored as $4(^-2x + 1)$ or $^-4(2x - 1)$. The preferred factorization is $^-4(2x - 1)$ because the coefficient of the first term of the binomial is positive.

EXERCISE 6.13.1

Factor each of the following so that the first term of the polynomial factor will have a positive coefficient.

1.  $^-5x + 15$                6.  $^-4x^2 - 8x + 24$
2.  $^-8x + 6$                 7.  $^-15x - 3$
3.  $^-10x - 5$                8.  $^-4x^2 - 11x + 9$
4.  $^-14x - 21$               9.  $^-6x^2 + 3x - 12$
5.  $^-7x + 9$                10.  $^-2x + 17$

## 6.14   Factoring where a Binomial is the Common Factor

$2x + 2$ is factored as $2(x + 1)$ because 2 is a common factor for the two terms of the polynomial.

$4x(x - 3) + 5(x - 3)$ can also be factored using the common factor method. $(x - 3)$ is a factor of $4x(x - 3)$ and $(x - 3)$ is also a factor of $5(x - 3)$. Hence $4x(x - 3) + 5(x - 3)$ is factored as $(x - 3)(4x + 5)$.

$$4x(x - 3) + 5(x - 3) = \underline{(x - 3)} \cdot 4x + \underline{(x - 3)} \cdot 5$$

$$= (x - 3)(4x + 5)$$

The expression $2x(x + 4) - (x + 4)$ can be factored because $(x + 4)$ is a common factor. The factorization is shown below. Note the introduction of $^-1$ in the second line of the example.

$$2x(x + 4) - (x + 4)$$

$$2x(x + 4) - 1(x + 4)$$

$$(x + 4)(2x - 1)$$

EXERCISE 6.14.1

Factor each of the following where the common factor is itself a binomial.

1.  $x(x + 8) - 3(x + 8)$      11.  $x(x + 8) + 4(x + 8)$
2.  $5x(x - 2) + 9(x - 2)$      12.  $x(x - 5) - 3(x - 5)$
3.  $x(x + 5) + 4(x + 5)$      13.  $x(x + 9) + 2(x + 9)$
4.  $x(x - 2) + 2(x - 2)$      14.  $x(2x + 1) - 5(2x + 1)$
5.  $4x(x + 6) - (x + 6)$      15.  $2x(x + 3) - 7(x + 3)$
6.  $x(x - 13) + (x - 13)$      16.  $x(x - 8) - (x - 8)$
7.  $2x(x + 3) - 5(x + 3)$      17.  $3x(x + 1) + (x + 1)$
8.  $x(x - 9) - 9(x - 9)$      18.  $x(3x - 5) - (3x - 5)$
9.  $2x(2x - 3) + 9(2x - 3)$      19.  $2x(x - 7) - 3(x - 7)$
10.  $x(x - 5) + 8(x - 5)$      20.  $x(2x + 5) + (2x + 5)$

## 6.15  Factoring by Parts

A useful skill to develop is the factoring of polynomials with four terms by the repeated use of the common factor method.

$x^2 + 5x + xy + 5y$ is a polynomial with four terms that can be factored into two binomials by factoring the first two terms, then the last two terms, and finally the entire polynomial.

$$x^2 + 5x + xy + 5y$$

$$(x^2 + 5x) + (xy + 5y)$$

$$x(x + 5) + y(x + 5)$$

$$(x + 5)(x + y)$$

$2x^2 - 3x + 10xy - 15y$ is another polynomial factorable in the same steps as used in the last example.

$$2x^2 - 3x + 10xy - 15y$$

$$(2x^2 - 3x) + (10xy - 15y)$$

$$x(2x - 3) + 5y(2x - 3)$$

$$(2x - 3)(x + 5y)$$

$x^2 - 5x - 2x + 10$ is a polynomial with four terms that could be simplified to a trinomial. (Why?) Nevertheless it can be factored in its present form by repeated use of the common factor method.

$$x^2 - 5x - 2x + 10$$

$$(x^2 - 5x) + (^-2x + 10)$$

$$x(x - 5) - 2(x - 5)$$

$$(x - 5)(x - 2)$$

Take special note of the manner in which the signs were handled in the preceding example. Because the third term was $^-2x$ it was necessary to factor the third and fourth terms as $^-2(x - 5)$. It is preferred to have the coefficient of the first term of the polynomial factor positive.

EXERCISE 6.15.1

Factor each of the following by repeated use of the common factor method.

| | | | |
|---|---|---|---|
| 1. | $x^2 - 6x + xy - 6y$ | 11. | $x^2 - 3x - x + 3$ |
| 2. | $x^2 - 5x - 2xy + 10y$ | 12. | $2x^2 + 6x - 3x - 9$ |
| 3. | $x^2 + 4x + xy + 4y$ | 13. | $4x^2 - 2x + 10x - 5$ |
| 4. | $x^2 - 9x - 3xy + 27y$ | 14. | $6x^2 - 21x + 2x - 7$ |
| 5. | $x^2 - 3x + 5xy - 15y$ | 15. | $5x^2 + 5x - 3x - 3$ |
| 6. | $2x^2 - 5x + 4xy - 10y$ | 16. | $6x^2 + 15x - 2x - 5$ |
| 7. | $2x^2 - 4x + xy - 2y$ | 17. | $12x^2 + 20x - 3x - 5$ |
| 8. | $x^2 + 7x - 3x - 21$ | 18. | $4x^2 + 6x + 6x + 9$ |
| 9. | $x^2 - 5x - 4x + 20$ | 19. | $100x^2 - 10x + 10x - 1$ |
| 10. | $x^2 + 9x - 2x - 18$ | 20. | $16x^2 - 20x - 20x + 25$ |

**6.16  Factoring Polynomials of the Form** $ax^2 + bx + c$

The last thirteen problems of the preceding exercise may have raised some questions in your mind because all of the polynomials could have been simplified to trinomials. If you had simplified $x^2 + 7x - 3x - 21$ to $x^2 + 4x - 21$ then you probably would have hidden its factors. In fact, our method for factoring polynomials such as $x^2 + 4x - 21$ and $2x^2 + 3x - 9$ is to rewrite the middle term as a binomial and, therefore, we end up factoring polynomials with four terms in exactly the same manner as required in the last exercise.

To factor $2x^2 + 3x - 9$ we follow the following procedure:

(1)   Multiply the first and last terms of the polynomial.

$$2x^2 \cdot {}^-9 = {}^-18x^2$$

(2)    Find factors of $^-18x^2$ which have the sum of $3x$, the middle term of the trinomial.

$6x \cdot {}^-3x = {}^-18x^2$

$6x + {}^-3x = 3x$

(3)    Rewrite the trinomial as a four-term polynomial using $6x + {}^-3x$ in place of the original middle term.

$2x^2 + 3x - 9$

$2x^2 + 6x - 3x - 9$

(4)    Factor the four-term polynomial by repeated use of the common factor method.

$2x^2 + 6x - 3x - 9$

$2x(x + 3) - 3(x + 3)$

$(x + 3)(2x - 3)$

(5)    Check the result by multiplying the factors.

$(x + 3)(2x - 3) = 2x^2 + 3x - 9$

To factor $9x^2 - 4$ we follow the same procedure as in the preceding example while recognizing that $9x^2 - 4$ is equivalent to the trinomial $9x^2 + 0x - 4$.

To factor $9x^2 - 4$ or $9x^2 + 0x - 4$:

(1)    Multiply the first and last terms.

$9x^2 \cdot {}^-4 = {}^-36x^2$

(2)    Find factors of $^-36x^2$ which have a sum of $0x$, the middle term of the trinomial.

$6x \cdot {}^-6x = {}^-36x^2$

$6x + {}^-6x = 0x$

(3)    Rewrite the polynomial with four terms.

$9x^2 - 4$

$9x^2 + 6x - 6x - 4$

(4)   Factor the four-term polynomial by the repeated use of the common factor method.

$9x^2 + 6x - 6x - 4$

$3x(3x + 2) - 2(3x + 2)$

$(3x + 2)(3x - 2)$

(5)   Check the result by multiplying the factors.

$(3x + 2)(3x - 2) = 9x^2 - 4$

The method for factoring polynomials explained in this section is more than just a direct way of finding factors; it is also possible to use the skill developed here to recognize when a polynomial has no binomial factors.

$x^2 + 5x + 3$ is called a prime polynomial because it can not be factored except by using 1 or $^-1$ as common factors.

Our factoring method shows $x^2 + 5x + 3$ to be prime in the following way. We follow the same procedure used earlier on the factorable polynomials.

(1)   $x^2 + 5x + 3$

Multiply the first and last terms.

$x^2 \cdot 3 = 3x^2$

(2)   Find factors of $3x^2$ which have a sum of $5x$. Since this is impossible, $x^2 + 5x + 3$ is a prime polynomial.

Whenever step (2) of the factoring method is impossible then the polynomial is prime. Whenever step (2) can be successfully completed then the polynomial is factorable.

EXERCISE 6.16.1

Factor.   If the polynomial is not factorable then state that it is prime.

| | | | |
|---|---|---|---|
| 1. | $2x^2 + 11x + 12$ | 11. | $x^2 + 2x - 15$ |
| 2. | $x^2 + 9x + 20$ | 12. | $x^2 - 5x - 24$ |
| 3. | $3x^2 - 2x - 5$ | 13. | $7x^2 - 8x + 1$ |
| 4. | $6x^2 + 13x + 2$ | 14. | $x^2 - 12x + 8$ |
| 5. | $x^2 - 2x - 35$ | 15. | $9x^2 - 3x - 2$ |
| 6. | $x^2 + 10x + 25$ | 16. | $6x^2 - 7x + 2$ |
| 7. | $3x^2 - 11x + 6$ | 17. | $3x^2 + 5x + 6$ |
| 8. | $x^2 - 5x + 6$ | 18. | $4x^2 + 11x + 6$ |
| 9. | $3x^2 - 10x + 3$ | 19. | $3x^2 - 8x + 4$ |
| 10. | $4x^2 - 4x - 3$ | 20. | $5x^2 - 12x + 4$ |

21. $2x^2 + 7x + 5$
22. $x^2 - 10x + 20$
23. $x^2 - 5x - 24$
24. $2x^2 + 7x + 6$
25. $3x^2 + 11x + 6$
26. $4x^2 - 16x + 7$
27. $5x^2 - 7x - 3$
28. $5x^2 - 8x - 5$
29. $3x^2 - 8x - 3$
30. $2x^2 - 5x - 12$
31. $x^2 - 9$
32. $x^2 - 49$
33. $4x^2 - 25$
34. $3x^2 - 14x + 8$
35. $8x^2 - 17x + 2$

36. $x^2 - 64$
37. $x^2 - 10$
38. $4x^2 - 12x + 9$
39. $9x^2 - 1$
40. $5x^2 - 11x + 2$
41. $x^2 - 1$
42. $x^2 - 81$
43. $25x^2 - 1$
44. $x^2 + 15x + 54$
45. $x^2 - 3x - 40$
46. $x^2 + 14x + 45$
47. $x^2 - 17x + 72$
48. $x^2 - 100$
49. $x^2 + 8x + 13$
50. $x^2 - 12x + 27$

### 6.17 Factoring Trinomials of the Form $x^2 + bx + c$

The factoring method used in the last exercise is satisfactory for factoring any trinomial of the form $ax^2 + bx + c$. It can be shortened to a much easier procedure for those trinomials of the form $x^2 + bx + c$, that is, where the coefficient of $x^2$ is 1.

To factor $x^2 + 5x - 14$:

(1) Find factors of the third term, $^-14$, that have a sum of 5, the coefficient of the middle term.

$$^-2 \cdot 7 = ^-14$$

$$^-2 + 7 = 5$$

(2) Use the numbers found in step (1) as second terms in binomials having $x$ as the first term.

$$(x - 2)(x + 7)$$

(3) Check by multiplying the factors.

$$(x - 2)(x + 7) = x^2 + 5x - 14$$

As another example, consider the following factoring of $x^2 - 64$.

(1) Think of $x^2 - 64$ as $x^2 + 0x - 64$. Find factors of $^-64$ that have a sum of zero.

$$8 \cdot {}^-8 = {}^-64$$

$$8 + {}^-8 = 0$$

(2)   Use the numbers found in step (1) as second terms in binomials having $x$ as the first term.

$(x + 8)(x - 8)$

(3)   Check by multiplying the factors.

$(x + 8)(x - 8) = x^2 - 64$

A polynomial such as $x^2 - 8x + 3$ can be shown to be prime because there are no factors of 3 that have a sum of $^-8$.

EXERCISE 6.17.1

Factor or state that the polynomial is prime.

| | |
|---|---|
| 1.  $x^2 + 7x + 6$ | 14.  $x^2 - 10x + 24$ |
| 2.  $x^2 + 2x + 1$ | 15.  $x^2 + 7x + 13$ |
| 3.  $x^2 - x - 20$ | 16.  $x^2 - 9$ |
| 4.  $x^2 - 12x + 20$ | 17.  $x^2 - 49$ |
| 5.  $x^2 + 13x - 30$ | 18.  $x^2 + 11x + 24$ |
| 6.  $x^2 - 2x + 1$ | 19.  $x^2 - 16$ |
| 7.  $x^2 - 5x + 8$ | 20.  $x^2 + 25$ |
| 8.  $x^2 - 9x + 18$ | 21.  $x^2 - 11x + 28$ |
| 9.  $x^2 - 5x - 24$ | 22.  $x^2 + 12x + 36$ |
| 10.  $x^2 - 8x + 12$ | 23.  $x^2 - 25$ |
| 11.  $x^2 - 18x + 81$ | 24.  $x^2 - 4$ |
| 12.  $x^2 + 8x + 7$ | 25.  $x^2 + 19x + 18$ |
| 13.  $x^2 - 10x + 21$ | |

## 6.18   Polynomials Requiring Repeated Factoring

The common factor method is basic to all the factoring of polynomials that has been presented in this text. When given a polynomial like $2x^2 - 12x - 14$ to factor, a student's first impulse may be toward a more difficult factoring method, but the first attempt should always be towards finding a common monomial factor for all the terms of the trinomial.

The best way to factor $2x^2 - 12x - 14$ is as follows:

(1)   $2x^2 - 12x - 14$ has 2 as a common factor. Hence it should first be factored as

$2(x^2 - 6x - 7)$

(2)   To complete the factoring of $2(x^2 - 6x - 7)$ we need only factor the trinomial.

$2(x^2 - 6x - 7)$

$2(x - 7)(x + 1)$

(3)   Check the factors by multiplying.

$$2[(x - 7)(x + 1)] = 2(x^2 - 6x - 7)$$
$$= 2x^2 - 12x - 14$$

The first attempt at factoring any polynomial should be to find a common monomial factor. This procedure will always lead to simpler factoring situations whenever the polynomial requires repeated factoring like the example shown above.

EXERCISE 6.18.1

Factor.   Each of the polynomials in this exercise can be factored in some way. Some require repeated factoring. Continue factoring until the polynomial factors are primes.

| | | | |
|---|---|---|---|
| 1. | $2x^2 - 16x + 14$ | 16. | $2x^2 + 8x + 8$ |
| 2. | $x^3 - 3x^2 + 8x$ | 17. | $5x - 15$ |
| 3. | $x^2 + 16x + 64$ | 18. | $x^3 - 81x$ |
| 4. | $x^3 + 3x^2 - 4x$ | 19. | $2x^2 - 8$ |
| 5. | $x^3 + 8x^2 - 9x$ | 20. | $10x^2 + 50x + 60$ |
| 6. | $3x^2 - 12$ | 21. | $x^2 - 11x + 24$ |
| 7. | $2x^2 + 8x - 24$ | 22. | $5x^2 - 10x + 40$ |
| 8. | $4x - 6$ | 23. | $3x^2 - 9x - 30$ |
| 9. | $x^4 - 36x^2$ | 24. | $4x^2 + 4x - 3$ |
| 10. | $5x^2 + 10x + 5$ | 25. | $6x^2 + 24x + 18$ |
| 11. | $2x^2 - 5x - 3$ | 26. | $4x^3 - 4x^2 - 80x$ |
| 12. | $5x^2 + 30x + 25$ | 27. | $x^3 - 5x^2 + 6x$ |
| 13. | $3x^3 + 6x^2 - 45x$ | 28. | $12x^2 - 8$ |
| 14. | $7x^2 - 7$ | 29. | $12x^2 - x - 1$ |
| 15. | $4x^2 - 12x - 40$ | 30. | $4x^2 - 9$ |

## 6.19   Simplifying Polynomial Fractions

Recall that any algebraic fraction of the form $\dfrac{x}{x}$ is equal to 1 where

it is understood that $x$ may not be replaced by zero. This allows us to simplify fractions such as $\dfrac{-2x^2y}{10xy^3}$ by dividing out the factors that the

denominator and numerator have in common.

$$\frac{-2x^2y}{10xy^3} = \frac{2xy}{2xy} \cdot \frac{-x}{5y^2} = \frac{-x}{5y^2}$$

Notice that we divide common *factors* out of fractions. We can not and must not divide out common terms unless they are also common factors.

For example, $\dfrac{x+7}{x-3}$ can not be simplified by dividing out the $x$'s simply

because $x$ is not a factor of either the numerator or denominator. Only common *factors* may be divided out of algebraic fractions.

To simplify $\dfrac{x^2-5x+4}{x^2-1}$ we must heed the warning of the last para-

graph and first factor both numerator and denominator.

$$\frac{x^2-5x+4}{x^2-1} = \frac{(x-1)(x-4)}{(x-1)(x+1)}$$

$$= \frac{x-4}{x+1}$$

The polynomial fraction $\dfrac{x^2-5x+4}{x^2-1}$ can be simplified to $\dfrac{x-4}{x+1}$   because

$(x-1)$ is a common factor for both numerator and denominator.

To simplify a polynomial fraction (1) factor both numerator and denominator and (2) divide out any common factors.


EXERCISE 6.19.1

Simplify:

1. $\dfrac{2x^2}{5x}$

2. $\dfrac{-4x^3y^2}{14xy}$

3. $\dfrac{x^2+5x+6}{x^2+7x+12}$

4. $\dfrac{x^2+3x-10}{x^2+6x-16}$

5. $\dfrac{3x - 6}{6x + 48}$        13. $\dfrac{x^2 + 5x - 14}{3x - 6}$

6. $\dfrac{x^2 + 5x}{x^2 + 12x + 35}$        14. $\dfrac{x^2 - 3x}{x^2 - 10x + 21}$

7. $\dfrac{x^2 - 13x + 30}{x^2 - 9}$        15. $\dfrac{3x - 24}{6x}$

8. $\dfrac{5x - 10}{4x - 8}$        16. $\dfrac{x^2 - 4}{3x + 6}$

9. $\dfrac{3x - 12}{x^2 - 7x + 12}$        17. $\dfrac{x^2 + 6x + 9}{2x + 6}$

10. $\dfrac{2x - 6}{4x + 4}$        18. $\dfrac{x^2 + 5x - 24}{x^2 + 10x + 16}$

11. $\dfrac{x^2 - 49}{x^2 + 8x + 7}$        19. $\dfrac{-3x^2 + 12x}{x^2 - 4x}$

12. $\dfrac{x^2 + 6x + 8}{x^2 + 3x + 2}$        20. $\dfrac{x^2 + 9x + 8}{x^2 - 64}$

## 6.20    The Multiplication of Polynomial Fractions

The multiplication problem $\dfrac{3}{4} \cdot \dfrac{5}{7}$ is completed by multiplying

numerators and multiplying denominators.

The multiplication problem $\dfrac{2x^2}{7} \cdot \dfrac{5}{3x}$ can be simplified by dividing out

the common factor, $x$, before multiplying numerators and multiplying
denominators.

$$\frac{2x^2}{7} \cdot \frac{5}{3x} = \frac{2x \cdot \cancel{x}}{7} \cdot \frac{5}{3 \cdot \cancel{x}} = \frac{10x}{21}$$

In multiplying polynomial fractions, the first step should always be the
factoring of each polynomial and the dividing out of any factors that are

common to both a numerator and a denominator. The multiplication is completed by multiplying numerators and multiplying denominators.

$$\frac{2x + 8}{x^2 - 7x + 12} \cdot \frac{x^2 - 9}{x^2 + 2x - 8}$$

$$\frac{2\cancel{(x + 4)}}{\cancel{(x - 3)}(x - 4)} \cdot \frac{(x + 3)\cancel{(x - 3)}}{(x - 2)\cancel{(x + 4)}}$$

$$\frac{2x + 6}{x^2 - 6x + 8}$$

Notice in the example that all polynomials were first factored. Then the common factors of $(x + 4)$ and $(x - 3)$ were divided out. Finally, the numerators were multiplied and the denominators were multiplied.

When two polynomials are identical except for having opposite signs for their like terms, then they have a quotient of $^{-}1$. This is shown in the following example.

$$\frac{x - 3}{5} \cdot \frac{7x}{3 - x}$$

$$\frac{x - 3}{5} \cdot \frac{7x}{^{-}1(x - 3)}$$

$$\frac{7x}{-5} \quad \text{or} \quad \frac{^{-}7x}{5}$$

EXERCISE 6.20.1

Multiply:

1. $\dfrac{3x^2}{5} \cdot \dfrac{4}{9x}$

2. $\dfrac{5x}{2y^2} \cdot \dfrac{4y}{7}$

3. $\dfrac{7x^2}{5} \cdot \dfrac{10}{3x^2}$

4. $\dfrac{3y^3}{7x} \cdot \dfrac{14x^2}{9y}$

5. $\dfrac{4x}{3y^2} \cdot \dfrac{6y}{7x^3}$

6. $\dfrac{x + 4}{5} \cdot \dfrac{7}{x + 4}$

7. $\dfrac{2x - 10}{x + 8} \cdot \dfrac{3}{x - 5}$

8. $\dfrac{x^2 + 3x - 4}{x + 3} \cdot \dfrac{4x + 12}{8x + 32}$

9. $\dfrac{x+2}{x-1} \cdot \dfrac{3x-3}{2x+8}$

20. $\dfrac{^{-}3x+18}{x^2-8x+12} \cdot \dfrac{x-2}{3x+12}$

10. $\dfrac{3x-6}{4} \cdot \dfrac{5}{x^2-4}$

21. $\dfrac{x^2-4}{x^2+8x+12} \cdot \dfrac{3x+18}{5x-10}$

11. $\dfrac{2}{5-2x} \cdot \dfrac{2x-5}{9y}$

22. $\dfrac{x^2-5x}{x^2+3x+2} \cdot \dfrac{x^2+7x+10}{x^2-25}$

12. $\dfrac{3x+12}{x^2-6x+5} \cdot \dfrac{x-5}{x+4}$

23. $\dfrac{x+7}{x^2+3x-28} \cdot \dfrac{x^2-16}{x^2+5x+4}$

13. $\dfrac{x+4}{x-2} \cdot \dfrac{5x-10}{x^2+6x+8}$

24. $\dfrac{4x^2}{x^2+4x-12} \cdot \dfrac{x^2-5x+6}{x^2-3x}$

14. $\dfrac{x-3}{2-x} \cdot \dfrac{x-2}{4}$

25. $\dfrac{x^2-36}{x^2+7x+6} \cdot \dfrac{3x+3}{x^2-8x+12}$

15. $\dfrac{x^2+x-12}{x-7} \cdot \dfrac{x-7}{3x-9}$

26. $\dfrac{2x^2+5x+3}{3x-6} \cdot \dfrac{4x-8}{4x+4}$

16. $\dfrac{x^2+7x+6}{x^2+4x+3} \cdot \dfrac{x+3}{x^2-36}$

27. $\dfrac{5x^2-15x+10}{x^2-4} \cdot \dfrac{3x+6}{10x}$

17. $\dfrac{x^2+10x+9}{x-7} \cdot \dfrac{x-7}{x+9}$

28. $\dfrac{4x^2-9}{x^2+3x} \cdot \dfrac{5x+15}{2x+3}$

18. $\dfrac{x^2+7x+12}{3x+12} \cdot \dfrac{3x-9}{x^2-9}$

29. $\dfrac{4x+12}{x^2-9} \cdot \dfrac{3-x}{8}$

19. $\dfrac{2-x}{x-3} \cdot \dfrac{x^2-9}{x-2}$

30. $\dfrac{x^2-4}{5x} \cdot \dfrac{10}{2-x}$

## 6.21  Adding Polynomial Fractions with the Same Denominator

The addition of fractions with the same denominator is accomplished by adding the numerators and keeping the same denominator. The procedure is justified by the Distributive Law of Multiplication over Addition.

In the addition problem $\dfrac{7}{x} + \dfrac{4}{x}$ the fractions have the same denomi-

nator, $x$. $\dfrac{1}{x}$ is a common factor for $\dfrac{7}{x}$ and $\dfrac{4}{x}$. This fact is used to justify the

addition shown below.

$$\frac{7}{x} + \frac{4}{x} = 7 \cdot \frac{1}{x} + 4 \cdot \frac{1}{x} = (7 + 4) \cdot \frac{1}{x} = 11 \cdot \frac{1}{x} = \frac{11}{x}$$

To add $\dfrac{x + 9}{x^2 - 6x - 11} + \dfrac{3x - 7}{x^2 - 6x - 11}$ we notice that the denomi-

nators are exactly the same. Therefore, the addition will be completed by
adding the numerators and retaining the same denominator.

$$\frac{x + 9}{x^2 - 6x - 11} + \frac{3x - 7}{x^2 - 6x - 11}$$

$$\frac{(x + 9) + (3x - 7)}{x^2 - 6x - 11}$$

$$\frac{4x + 2}{x^2 - 6x - 11}$$

The fractions of $\dfrac{x^2 - 6}{x - 1} - \dfrac{2x - 3}{x - 1}$ are separated by a minus sign.

Since the minus sign means addition of the opposite, the problem is completed in the following manner.

$$\frac{x^2 - 6}{x - 1} - \frac{2x - 3}{x - 1}$$

$$\frac{(x^2 - 6) - (2x - 3)}{x - 1}$$

$$\frac{x^2 - 6 - 2x + 3}{x - 1}$$

$$\frac{x^2 - 2x - 3}{x - 1}$$

234    *Algebraic Polynomials, Factoring, and Fractions*

EXERCISE 6.21.1

Add:

1. $\dfrac{6}{5y} + \dfrac{7}{5y}$

11. $\dfrac{4x + 7}{x^2 + 9x - 1} + \dfrac{3x - 8}{x^2 + 9x - 1}$

2. $\dfrac{4x}{x + 1} + \dfrac{3x - 4}{x + 1}$

12. $\dfrac{x^2 - 3x + 2}{x - 6} - \dfrac{x^2 + 3x + 2}{x - 6}$

3. $\dfrac{2x - 1}{x + 3} - \dfrac{x + 2}{x + 3}$

13. $\dfrac{3x + 7}{x^2 - 80} + \dfrac{5x - 11}{x^2 - 80}$

4. $\dfrac{x^2 - 3}{x + 9} + \dfrac{x^2 - 3x - 2}{x + 9}$

14. $\dfrac{x^2 + 5x + 3}{x^2 - 11x + 7} + \dfrac{2x^2 - 16}{x^2 - 11x + 7}$

5. $\dfrac{x + 4}{x^2 - 6x + 1} - \dfrac{3x - 2}{x^2 - 6x + 1}$

15. $\dfrac{x^2 - 5x + 7}{x - 11} - \dfrac{x^2 - 9x - 3}{x - 11}$

6. $\dfrac{x + 5}{x^2 - 9x + 3} - \dfrac{x^2 - 3x + 2}{x^2 - 9x + 3}$

16. $\dfrac{x^2 - 3x + 5}{x^2 - 11} + \dfrac{2x^2 - 1}{x^2 - 11}$

7. $\dfrac{x^2 + 3x + 2}{x + 7} + \dfrac{x - 3}{x + 7}$

17. $\dfrac{2x - 3y}{x^2 + 7x + 9} - \dfrac{5x + 2y}{x^2 + 7x + 9}$

8. $\dfrac{x^2 - 3}{x + 9} - \dfrac{2x + 7}{x + 9}$

18. $\dfrac{x^2 - 3}{4y^3 - 1} - \dfrac{2x - 3}{4y^3 - 1}$

9. $\dfrac{x^2 + 3x - 5}{2x} + \dfrac{3x - 6}{2x}$

19. $\dfrac{6x^2 - 3y}{2x + y} + \dfrac{5x + 2y}{2x + y}$

10. $\dfrac{8x - 5}{x + 9} - \dfrac{3x + 1}{x + 9}$

20. $\dfrac{4x + 1}{x^2 + 2} - \dfrac{2x - 3}{x^2 + 2}$

## 6.22  Finding Least Common Multiples for Pairs of Polynomials

The addition of polynomial fractions with different denominators is usually easier when a common denominator is found. In this section a method for finding least common multiples is shown. The least common multiples found in this way will be used as common denominators in the next section on adding polynomial fractions.

The least common multiple for 4 and 6 is 12 because 12 is the smallest counting number having both 4 and 6 as factors. The student probably recognizes 12 as the least common multiple of 4 and 6 because of his knowledge of the basic multiplication facts, but a method for finding least common multiples must be developed for numbers such as 28 and 49 or polynomials such as $x^2 - 9x + 14$ and $x^2 + 2x - 8$.

To find the least common multiple of 28 and 49.

(1)   Find the highest common factor of 28 and 49. This is 7.

(2)   Divide the product $28 \cdot 49$ by the common factor 7.

$$\frac{28 \cdot 49}{7} = \frac{4 \cdot 49}{1} = \frac{196}{1} = 196$$

196 is the least common multiple of 28 and 49.

To find the least common multiple of $x^2 - 9x + 14$ and $x^2 + 2x - 8$:

(1)   Find the common factor for $x^2 - 9x + 14$ and $x^2 + 2x - 8$.

$$x^2 - 9x + 14 = (x - 2)(x - 7)$$

$$x^2 + 2x - 8 = (x + 4)(x - 2)$$

$(x - 2)$ is the common factor.

(2)   Divide the product of $x^2 - 9x + 14$ and $x^2 + 2x - 8$ by the common factor $(x - 2)$.

$$\frac{(x^2 - 9x + 14)(x^2 + 2x - 8)}{x - 2} = \frac{(x - 2)(x - 7) \cdot (x + 4)(x - 2)}{x - 2}$$

$(x - 7)(x + 4)(x - 2)$ is the least common multiple of $x^2 - 9x + 14$ and $x^2 + 2x - 8$.


EXERCISE 6.22.1

Find the least common multiple for each pair. Leave the polynomials in factored form.

1.   16   and   12
2.   20   and   25

3. $35$ and $63$
4. $33$ and $55$
5. $5x^2$ and $10x$
6. $2xy$ and $5y$
7. $x^2 + 10x - 11$ and $5x - 5$
8. $x^2 + 7x + 6$ and $3x + 3$
9. $x^2 - 49$ and $5x - 35$
10. $x^2 + 5x + 4$ and $6x + 24$
11. $x^2 - 7x - 18$ and $x^2 - 10x + 9$
12. $x^2 - 64$ and $x^2 - x - 56$
13. $x^2 + x - 12$ and $x^2 - 5x + 6$
14. $x^2 - x - 56$ and $x^2 - 9x + 8$
15. $2x^2 - 7x + 6$ and $8x - 12$
16. $x^2 + 4x - 21$ and $x^2 - 49$
17. $4x^2 - 4x + 1$ and $2x^2 + 9x - 5$
18. $x^2 + 7x + 6$ and $x^2 + 9x + 18$
19. $x^2 - 1$ and $x^2 + 3x - 4$
20. $x^2 + 16x + 63$ and $x^2 + x - 72$

## 6.23 Adding Polynomial Fractions

Any fraction may be multiplied by 1 without changing its value. When 1 is written as $\dfrac{5}{5}$, $\dfrac{x-3}{x-3}$, and $\dfrac{2x+5}{2x+5}$ then multiplying a fraction by it may appear to change the value of the fraction, but obviously it can not.

$$\frac{4}{7} = \frac{4}{7} \cdot \frac{5}{5} = \frac{20}{35}$$

$$\frac{x-1}{x+3} = \frac{x-1}{x+3} \cdot \frac{x-3}{x-3} = \frac{x^2 - 4x + 3}{x^2 - 9}$$

$$\frac{x-7}{(x-2)(x+6)} = \frac{x-7}{(x-2)(x+6)} \cdot \frac{2x+5}{2x+5} = \frac{2x^2 - 9x - 35}{(x-2)(x+6)(2x+5)}$$

The three pairs of equal fractions above show situations in which the denominators have been changed, but the value of the fractions remained the same. To add two polynomial fractions a common denominator is needed and the procedure above indicates the method by which the denominator can be changed without altering the value of the fraction.

The addition of $\dfrac{x+5}{4x} + \dfrac{x-1}{2x^2}$ is shown below.

(1)  $\dfrac{x+5}{4x} + \dfrac{x-1}{2x^2}$  First, find the least common multiple of the denominators.

$\dfrac{4x \cdot 2x^2}{2x} = 4x^2$

(2)  Multiply $\dfrac{x+5}{4x}$ by $\dfrac{x}{x}$ so that its denominator will be $4x^2$.

$\dfrac{x+5}{4x} \cdot \dfrac{x}{x} = \dfrac{x^2+5x}{4x^2}$

(3)  Multiply $\dfrac{x-1}{2x^2}$ by $\dfrac{2}{2}$ so that its denominator will be $4x^2$.

$\dfrac{x-1}{2x^2} \cdot \dfrac{2}{2} = \dfrac{2x-2}{4x^2}$

(4)  Complete the addition by adding the numerators now that the denominators are the same.

$\dfrac{x^2+5x}{4x^2} + \dfrac{2x-2}{4x^2} = \dfrac{x^2+7x-2}{4x^2}$

The same four steps are used to add

$$\dfrac{x+4}{x^2-5x-24} - \dfrac{2x-1}{x^2-9}$$

(1)  The least common multiple of the denominators is

$\dfrac{(x-8)(x+3)\cdot(x+3)(x-3)}{x+3} = (x-8)(x+3)(x-3)$

(2)  $\dfrac{x+4}{(x-8)(x+3)} \cdot \dfrac{x-3}{x-3} = \dfrac{x^2+x-12}{(x-8)(x+3)(x-3)}$

(3)  $\dfrac{2x-1}{(x+3)(x-3)} \cdot \dfrac{x-8}{x-8} = \dfrac{2x^2-17x+8}{(x+3)(x-3)(x-8)}$

(4) $\dfrac{x + 4}{x^2 - 5x - 24} - \dfrac{2x - 1}{x^2 - 9}$

$$\dfrac{x^2 + x - 12}{(x - 8)(x + 3)(x - 3)} - \dfrac{2x^2 - 17x + 8}{(x + 3)(x - 3)(x - 8)}$$

$$\dfrac{(x^2 + x - 12) - (2x^2 - 17x + 8)}{(x - 8)(x + 3)(x - 3)}$$

$$\dfrac{{}^{-}x^2 + 18x - 20}{(x - 8)(x + 3)(x - 3)}$$

EXERCISE 6.23.1

Add:

1. $\dfrac{5}{3x^2} + \dfrac{7}{6x}$

2. $\dfrac{4}{5x^2} - \dfrac{3}{2x^2}$

3. $\dfrac{x - 3}{6x} + \dfrac{3}{8x^2}$

4. $\dfrac{x - 7}{15} + \dfrac{x + 2}{10}$

5. $\dfrac{x - 6}{3x} - \dfrac{x + 1}{4x^2}$

6. $\dfrac{5}{x^2 + 10x + 9} + \dfrac{3}{x^2 - 1}$

7. $\dfrac{4}{x^2 + 3x + 2} + \dfrac{7}{x^2 + 5x + 6}$

8. $\dfrac{x + 3}{x^2 - 7x + 12} - \dfrac{x - 2}{x^2 - 3x - 4}$

9. $\dfrac{x+4}{x^2-2x-15} + \dfrac{x-5}{x^2+7x+12}$

10. $\dfrac{x+1}{x^2-4x+3} - \dfrac{x-2}{x^2+x-2}$

11. $\dfrac{x+7}{x^2-3x+2} - \dfrac{x+2}{x^2-5x+6}$

12. $\dfrac{x+1}{x^2-4} - \dfrac{x+2}{x^2+3x-10}$

13. $\dfrac{x-6}{2x^2-5x-3} + \dfrac{x+2}{2x^2-x-1}$

14. $\dfrac{x+3}{2x^2-7x-15} - \dfrac{x-2}{2x^2+5x+3}$

15. $\dfrac{5}{x^2-9} - \dfrac{3}{x^2-2x-3}$

16. $\dfrac{x-1}{x^2+5x+6} - \dfrac{x-2}{x^2+2x-3}$

17. $\dfrac{2x-1}{3x} + \dfrac{x+7}{4x}$

18. $\dfrac{2x-1}{x^2-x-2} + \dfrac{x+3}{x^2-5x+6}$

19. $\dfrac{x-5}{2x^2} - \dfrac{x+2}{6x}$

20. $\dfrac{x+4}{x^2+6x+5} - \dfrac{x-2}{x^2-4x-5}$

# 7 SOLVING EQUATIONS AND INEQUALITIES

## 7.1 Introduction

In previous chapters, notably chapters 2, 3, and 4, the solution of linear equations and linear inequalities has been studied. In this chapter we return to the solution of equations and inequalities by offering a review of the previous work on linear equations and inequalities, extending these skills to develop methods for solving equations involving polynomial fractions, and introducing a new type of equation and inequality called the quadratic.

Recall that an equation or an inequality is an open sentence that is neither true nor false. This is because an open sentence contains a variable which is used like a blank space; in other words, the open sentence is not a statement until a replacement for the variable is chosen. Two examples of open sentences are shown below.

$$3x + 5 = 19$$

$$x \text{ is a U.S. Senator.}$$

The first of these open sentences is the type that we call an equation. The second example of an open sentence is offered here to emphasize the use of variables in mathematics. Often we hear that for the open sentence $x + 3 = 5$ then $x$ is 2. If taken literally this is nonsense. $x$ is a letter of the alphabet that is being used as a variable; $x$ is definitely not 2. What is meant is "If $x$ is replaced by 2 in the open sentence $x + 3 = 5$ then a true statement, $2 + 3 = 5$, will result." Variables are not numbers; they represent places to be filled, but they do not determine their own replacements. 4 is a perfectly good replacement for $x$ in the inequality $x + 2 > 9$; it provides the statement $4 + 2 > 9$ which happens to be false, but the role of the variable does not demand only those replacements that will result in true statements.

Students are often asked to find those replacements of a variable that will result in true statements, but this is another requirement that has been placed upon the open sentence rather than any inherent necessity placed upon the problem by its variable.

Whenever a variable appears in an open sentence it must be accompanied by a stated or implied replacement set, called the domain. Only elements of the replacement set are allowed as replacements for the variable. Oftentimes the replacement set to be used with an open sentence will play a crucial role in arriving at truth sets. Below is shown a quadratic equation of the type that will be solved toward the end of this chapter. Although you may not know how the elements of the truth sets were obtained you should observe that the truth sets are different and the difference is because of the replacement set used.

$$2x^2 + 5x - 3 = 0$$

(1)   Using the set of counting numbers $\{1, 2, 3, \ldots\}$ as the replacement set, the equation $2x^2 + 5x - 3 = 0$ has the empty set, $\{\quad\}$, as its truth set because there is no counting number that will result in a true statement.

(2)   Using the set of integers $\{\ldots, {}^-2, {}^-1, 0, 1, 2, \ldots\}$ as the replacement set, the equation $2x^2 + 5x - 3 = 0$ has the singleton set $\{{}^-3\}$, as its truth set because $2 \cdot ({}^-3)^2 + 5 \cdot {}^-3 - 3 = 0$ is a true statement.

(3)   Using the set of rational numbers as the replacement set, the equation $2x^2 + 5x - 3 = 0$ has a doubleton set. $\left\{\dfrac{1}{2}, {}^-3\right\}$ , because $2 \cdot \left(\dfrac{1}{2}\right)^2 + 5 \cdot \dfrac{1}{2} - 3 = 0$ is a true statement.

Hence we see that $2x^2 + 5x - 3 = 0$ can have $\{\ \ \}$, $\{-3\}$, or $\left\{\dfrac{1}{2}, -3\right\}$ as its truth set depending on what set is used as the replacement set for the variable. Every equation or inequality is solved with respect to a particular set of numbers that serves as the replacement set for the variable.

## 7.2 Solving Open Sentences Using the Set of Counting Numbers

We begin our review of the solution of linear equalities and inequalities by studying open sentences that have the set of counting numbers $\{1, 2, 3, \ldots\}$ as the replacement set. This might be called a primitive method for developing solutions because it involves only the procedure of simplifying algebraic expressions and an understanding of the use of variables. The value of solving open sentences using $\{1, 2, 3, \ldots\}$ as the replacement set is that this procedure emphasizes the primary aim of finding truth sets rather than concealing this aim in a discussion of procedures.

To solve the open sentence $x + 9 > 12$ using $\{1, 2, 3, \ldots\}$ is simply to find those replacements of $x$ that will result in a true statement when used in $x + 9 > 12$. No procedure is really necessary for this task. If it is not already obvious that $\{4, 5, 6, \ldots\}$ is the truth set then the student can test various elements of $\{1, 2, 3, \ldots\}$ until he discovers it.

To solve $3x + 13 = 34$ using $\{1, 2, 3, \ldots\}$ is simply to find any counting number replacement(s) for $x$ so that $3x + 13 = 34$ will become a true statement. The truth set here may not be obvious, but it can easily be found by testing the equation with a few counting numbers. Suppose $x$ is replaced by 5; $3 \cdot 5 + 13 = 34$ is false, but since $3 \cdot 5 + 13 < 34$ we should note that any counting number less than 5 will also make $3x + 13 = 34$ a false statement. Suppose $x$ is replaced by 10; $3 \cdot 10 + 13 = 34$ is also false, but since $3 \cdot 10 + 13 > 34$ we can eliminate any values greater than 10. (Why?) This leaves only 6, 7, 8, and 9 to be tested and should lead to the truth set $\{7\}$.

To solve $3(x + 7) + 5(x + 2) < 4 + 5(2 + 7)$ using $\{1, 2, 3, \ldots\}$ is simply to find any counting number replacements for $x$ that will make $3(x + 7) + 5(x + 2) < 4 + 5(2 + 7)$ a true statement. This problem emphasizes the one procedural step involved in solving any open sentence. It is unnecessarily cumbersome to use the inequality $3(x + 7) + 5(x + 2) < 4 + 5(2 + 7)$ when the equivalent inequality $8x + 31 < 49$ would be far easier to use in testing various counting number replacements. The first and only procedural step to be followed in solving open sentences using the counting numbers is to always simplify the two members of the open sentence to obtain the simplest equivalent open sentence possible.

Hence,     $3(x + 7) + 5(x + 2) < 4 + 5(2 + 7)$

$$3x + 21 + 5x + 10 < 4 + 5 \cdot 9$$

$$8x + 31 < 4 + 45$$

$$8x + 31 < 49$$

Testing various counting numbers in $8x + 31 < 49$ results in finding the truth set of $\{1, 2\}$.


EXERCISE 7.2.1

Use the set of counting numbers, $\{1, 2, 3, \ldots\}$, to find the truth sets for each of the following.

| | |
|---|---|
| 1. $x + 5 = 12$ | 10. $6x + 5 = 43$ |
| 2. $7x = 42$ | 11. $4x + 1 > 19$ |
| 3. $x + 14 < 20$ | 12. $2x + 3 < 1$ |
| 4. $x + 11 > 35$ | 13. $(2x + 7) + 2 = 29$ |
| 5. $x + 7 = 91$ | 14. $4x + (3x + 1) = 50$ |
| 6. $5x = 62$ | 15. $5(x + 2) + 3 = 43$ |
| 7. $x + 43 < 97$ | 16. $2x + 4(x + 5) = 44$ |
| 8. $x + 45 > 106$ | 17. $3x + 7(x + 2) > 9$ |
| 9. $2x + 3 = 19$ | |

18. $5(x + 1) + 2(x + 4) < 71$
19. $7(x + 3) + 5(x + 2) > 7 + 9(6 + 3)$
20. $4x + 4(2x + 3) + 7 < 7 \cdot 4 + 3 \cdot 6$


### 7.3  Solving Open Sentences Using the Set of Integers

For equations such as $x + 9 = 13$ and $4x = 28$ there is little, if any, difficulty in finding the truth set for any given replacement set. No procedure, no steps, no directions are really necessary, nor desirable, for such equations because the truth set should be obvious without any manipulation of the numbers and symbols.

The equations $7x + 13 = 48 + 2x$ and $2x + 17 = 5x + 8$ probably do not have obvious truth sets. Although these equations could be solved by testing various elements of the replacement set, a method or procedure for finding truth sets now becomes more desirable. In developing such a method the set of integers plays a vital role—not only in providing additional elements as possible candidates for truth sets but also because the set of integers enjoys a particular property that is not shared by the counting numbers.

Every element in the set of integers $\{\ldots, {}^-2, {}^-1, 0, 1, 2, \ldots\}$ has an opposite. This means that for any integer whatsoever there is another integer that can be added to it to give the sum of zero. Since zero is the identity element for addition, the fact that every integer has an opposite allows us to eliminate any term desired from one member of an equation or inequality.

Given the equation $2x + 17 = 5x + 8$ and using $\{\ldots, {}^-2, {}^-1, 0, 1, 2, \ldots\}$ as the replacement set, the method of solving depends upon finding other, simpler equations which have the same truth set. When the truth set becomes obvious for some equivalent equation then it is also obvious for $2x + 17 = 5x + 8$.

$$2x + 17 = 5x + 8$$

(1)   Add $^-17$ to both members of the equation to eliminate the 17 from the left member.

$$2x + 17 = 5x + 8$$
$$(2x + 17) + {}^-17 = (5x + 8) + {}^-17$$
$$2x = 5x - 9$$

(2)   Add $^-5x$ to both members of the equation to eliminate the $5x$ from the right member.

$$2x = 5x - 9$$
$$2x + {}^-5x = (5x - 9) + {}^-5x$$
$$^-3x = {}^-9$$

(3)   The truth set of $^-3x = {}^-9$ is obviously $\{3\}$. Hence the truth set of $2x + 17 = 5x + 8$ is also $\{3\}$.

The justification for adding $^-17$ and $^-5x$ to both members of the equation of the preceding example is called the Additive Property of Equality and Inequalities.

*Additive Property of Equality and Inequalities*

Any number or algebraic expression may be added to both members of an open sentence to obtain another open sentence with the same truth set.

Given below is a procedure for solving open sentences, but remember that whenever the truth set is obvious no further work is necessary or desirable.

The procedure that may be followed in solving any open sentence using the set of integers as the replacement set is

(1)    Simplify both members of the open sentence.

(2)    By adding opposites eliminate all terms containing the variable from one member of the open sentence and all terms not containing the variable from the other member of the open sentence.

(3)    Test the resulting open sentence using various integers. Generally the truth set will be obvious.

EXERCISE 7.3.1

Find truth sets using the set of integers $\{\ldots, {}^-2, {}^-1, 0, 1, 2, \ldots\}$ as the replacement set.

| | |
|---|---|
| 1. $x + 19 = 6$ | 7. $6x + 13 = 1$ |
| 2. $^-7x = 28$ | 8. $9x + 5 = 37$ |
| 3. $x + 3 > {}^-1$ | 9. $2x - 7 = 5x - 10$ |
| 4. $4x < 8$ | 10. $9x + 8 = 2x - 13$ |
| 5. $3x + 7 = 43$ | 11. $4x - 11 = 6x + 5$ |
| 6. $5x - 6 = 34$ | 12. $5x + 3 > 2x + 1$ |

13.    $2x - 3 < 12 - x$
14.    $3x - (4x + 1) = 6 - 2x$
15.    $3(5x - 2) + x = 10x + 3$
16.    $2(x - 7) - 3(x + 1) = 1$
17.    $4(x + 5) = 8 - 2(x - 3)$
18.    $4(2x - 7) + 5 = 2x + 3(x + 1)$
19.    $8 + 3(x + 2) > 7$
20.    $9 + 4(2x - 3) < 8$

## 7.4    Solving Open Sentences Using the Set of Rational Numbers

The truth set of $\dfrac{3}{4}x - \dfrac{7}{5} = \dfrac{3}{2}$ using the set of rational numbers as the replacement set is hardly obvious. In fact, probably the most obvious fact about the equation is that a method or procedure is necessary to find its truth set.

The set of rational numbers provides a powerful set of numbers for attacking equations such as $\dfrac{3}{4}x - \dfrac{7}{5} = \dfrac{3}{2}$. Like the integers, every rational number has an opposite. Unlike the integers, every rational number except zero also has a reciprocal. Given any nonzero rational number there is another rational number so that their product is one. Since one is the identity element for multiplication, the fact that every nonzero rational number has a reciprocal allows us to always generate an open sentence in which the variable is isolated in one member of the equation or inequality.

Since equations such as $x = \dfrac{^-13}{8}$ and $x = \dfrac{5}{42}$ have obvious truth sets, the existence of opposites and reciprocals in the set of rational numbers is tantamount to insuring obvious solutions for any linear equations of the form $ax + b = c$ where the rational numbers are used as the domain of the variable and $a \neq 0$.

The truth set of $\dfrac{3}{4} x - \dfrac{7}{5} = \dfrac{3}{2}$ may be obtained in the following steps:

(1)  Add $\dfrac{7}{5}$ to both members of the equality.

$$\frac{3}{4} x - \frac{7}{5} = \frac{3}{2}$$

$$\left(\frac{3}{4} x - \frac{7}{5}\right) + \frac{7}{5} = \frac{3}{2} + \frac{7}{5}$$

$$\frac{3}{4} x = \frac{29}{10}$$

(2)  Multiply both members of the equation by $\dfrac{4}{3}$, the reciprocal of $\dfrac{3}{4}$.

$$\frac{3}{4} x = \frac{29}{10}$$

$$\frac{4}{3}\left(\frac{3}{4} x\right) = \frac{4}{3} \cdot \frac{29}{10}$$

$$x = \frac{58}{15}$$

Using the set of rational numbers as the domain, $\dfrac{3}{4}x - \dfrac{7}{5} = \dfrac{3}{2}$ has the

truth set $\left\{\dfrac{58}{15}\right\}$ because the truth set of $x = \dfrac{58}{15}$ is obviously $\left\{\dfrac{58}{15}\right\}$.

The justification of the multiplication of both members of the equation

by $\dfrac{4}{3}$ is called the Multiplicative Property of Equality.

*Multiplicative Property of Equality*

> Both members of an equation may be multiplied by any nonzero number to obtain another equation with the same truth set.

EXERCISE 7.4.1

Use the set of rational numbers as the domain and find truth sets for:

1. $x - \dfrac{1}{2} = \dfrac{3}{4}$

2. $x + \dfrac{3}{5} = \dfrac{7}{10}$

3. $x + \dfrac{2}{3} = \dfrac{1}{6}$

4. $\dfrac{2}{5}x = \dfrac{7}{3}$

5. $\dfrac{^{-}10}{9}x = \dfrac{7}{4}$

6. $4x = {}^{-}5$

7. $\dfrac{2}{3}x + \dfrac{5}{6} = \dfrac{1}{2}$

8. $\dfrac{5}{6} - \dfrac{1}{3}x = \dfrac{1}{4}$

9. $\dfrac{3}{4}x + \dfrac{2}{3} = \dfrac{1}{5}$

10. $\dfrac{4}{3}x - \dfrac{3}{4} = \dfrac{1}{2}$

11. $2x - \dfrac{2}{5} = \dfrac{9}{4}x + \dfrac{1}{3}$

12. $3x - \dfrac{1}{2} = \dfrac{5}{8} + \dfrac{15}{4}x$

13. $x - 3 = \dfrac{1}{5} - \dfrac{2}{5}x$

14. $\dfrac{7}{8}x - 4 = 2 - \dfrac{3}{4}x$

15. $\dfrac{2}{7}x - \dfrac{3}{4} = x + 2$

16. $\dfrac{3}{5}x + 5 = \dfrac{1}{2}x + 1$

17. $2x + \dfrac{3}{7} = \dfrac{1}{5}x + \dfrac{4}{7}$

19. $4x + 5 = \dfrac{13}{2} + \dfrac{5}{2}x$

18. $\dfrac{3}{4}x - \dfrac{1}{2} = x + \dfrac{1}{3}$

20. $\dfrac{2}{3}x - \dfrac{2}{5} = 2 - \dfrac{1}{5}x$

There are two differences that should be noted between the Multiplicative Property of Equality and the Additive Property of Equalities and Inequalities given earlier. First, note that the Additive Property applies to any equality or inequality while the Multiplicative Property as given above only applies to equations. Later in this section a Multiplicative Property for Inequalities will be given. The second difference to be noted is that the Multiplicative Property only applies to the use of nonzero numbers whereas the Additive Property may be used with any number or algebraic expression. This difference will be especially important later in the section on equations involving polynomial fractions.

The inequality $\dfrac{2}{3}x + \dfrac{5}{6} < \dfrac{1}{4}$ is simplified as follows:

(1) Add $\dfrac{^-5}{6}$ to both members of the inequality.

$$\frac{2}{3}x + \frac{5}{6} < \frac{1}{4}$$

$$\left(\frac{2}{3}x + \frac{5}{6}\right) + \frac{^-5}{6} < \frac{1}{4} + \frac{^-5}{6}$$

$$\frac{2}{3}x < \frac{^-7}{12}$$

(2) Multiply both members by $\dfrac{3}{2}$.

$$\frac{2}{3}x < \frac{^-7}{12}$$

$$\frac{3}{2} \cdot \frac{2}{3}x < \frac{3}{2} \cdot \frac{^-7}{12}$$

$$x < \frac{^-7}{8}$$

The preceding two steps show that $\frac{2}{3}x + \frac{5}{6} < \frac{1}{4}$ may be simplified to

$x < \frac{-7}{8}$. The two inequalities are equivalent because they have the same

truth set.

The inequality $\frac{1}{3}x < \frac{5}{4}x + \frac{4}{7}$ may be simplified as follows:

(1)   Add $\frac{-5}{4}x$ to both members of the inequality.

$$\frac{1}{3}x < \frac{5}{4}x + \frac{4}{7}$$

$$\frac{1}{3}x + \frac{-5}{4}x < \left(\frac{5}{4}x + \frac{4}{7}\right) + \frac{-5}{4}x$$

$$\frac{-11}{12}x < \frac{4}{7}$$

(2)   Multiply both members of the inequality by $\frac{-12}{11}$ *and reverse*

*the direction of the inequality.*

$$\frac{-11}{12}x < \frac{4}{7}$$

$$\frac{-12}{11} \cdot \frac{-11}{12}x > \frac{-12}{11} \cdot \frac{4}{7}$$

$$x > \frac{-48}{77}$$

$\frac{1}{3}x < \frac{5}{4}x + \frac{4}{7}$ can be simplified to $x > \frac{-48}{77}$. The two inequalities are

equivalent.

Special attention should be given to the second step of the two preceding examples since they illustrate the following Multiplicative Property of Inequality.

*Multiplicative Property of Inequality*

Both members of an inequality may be multiplied by any positive number to obtain another equivalent inequality. If both members of an inequality are multiplied by a negative number, then the direction of the inequality must be reversed to obtain an equivalent inequality.

EXERCISE 7.4.2

Find simpler, equivalent inequalities for:

1. $x - \dfrac{5}{6} < \dfrac{4}{7}$

2. $\dfrac{4}{3} - x < \dfrac{9}{10}$

3. $x + \dfrac{5}{4} > \dfrac{-1}{2}$

4. $\dfrac{2}{3}x < \dfrac{4}{9}$

5. $\dfrac{-5}{7}x > \dfrac{2}{5}$

6. $\dfrac{3}{5}x - \dfrac{1}{3} < \dfrac{4}{5}$

7. $\dfrac{1}{4} - \dfrac{8}{9}x > \dfrac{5}{9}$

8. $4x - \dfrac{1}{3} < \dfrac{6}{5}$

9. $x + \dfrac{4}{7} < \dfrac{5}{4}x - \dfrac{1}{3}$

10. $\dfrac{3}{4}x - 2 > 5 - \dfrac{1}{2}x$

11. $x + \dfrac{5}{7} > \dfrac{2}{3}$

12. $\dfrac{2}{5} - x < 3$

13. $x - \dfrac{1}{4} > \dfrac{1}{5}$

14. $\dfrac{3}{4}x < \dfrac{-1}{4}$

15. $\dfrac{-5}{6}x > \dfrac{1}{2}$

16. $\dfrac{2}{5}x - \dfrac{1}{3} < 2$

17. $2 - \dfrac{1}{5}x > 5$

18. $3 - 2x < \dfrac{1}{2}x - 1$

19. $8x - 3 > 5x - \dfrac{7}{2}$

20. $x + \dfrac{7}{8} < \dfrac{1}{2} - \dfrac{1}{2}x$

### 7.5   An Alternate Method for Solving Fractional Equations

In previous sections on solving equations like $\frac{5}{7}x - \frac{8}{11} = \frac{1}{2}$ it has always been suggested that the first step be the addition of opposites to obtain an equivalent equation in which both members are monomials. The second step was multiplication by the reciprocal of the coefficient of the variable.

$$\frac{5}{7}x - \frac{8}{11} = \frac{1}{2}$$

(1)   First, add $\frac{8}{11}$.

$$\left(\frac{5}{7}x - \frac{8}{11}\right) + \frac{8}{11} = \frac{1}{2} + \frac{8}{11}$$

$$\frac{5}{7}x = \frac{27}{22}$$

(2)   Second, multiply by $\frac{7}{5}$.

$$\frac{7}{5} \cdot \frac{5}{7}x = \frac{7}{5} \cdot \frac{27}{22}$$

$$x = \frac{189}{110}$$

The addition need not always precede any multiplication. In fact, at times it may be much easier to first multiply both members of an equation by a well-selected number. In this section such a procedure is explained.

The equation $\frac{2}{3}x + \frac{1}{5} = \frac{4}{7}$ has 3, 5, and 7 as denominators. The Multiplicative Property of Equality allows both members of an equation to be multiplied by any nonzero number. What advantages may be gained by multiplying both members of $\frac{2}{3}x + \frac{1}{5} = \frac{4}{7}$ by 3, 5, 7, or all three of them?

$$\frac{2}{3}x + \frac{1}{5} = \frac{4}{7}$$

$$3\left(\frac{2}{3}x + \frac{1}{5}\right) = 3 \cdot \frac{4}{7}$$

$$2x + \frac{3}{5} = \frac{12}{7}$$

In what way might $2x + \dfrac{3}{5} = \dfrac{12}{7}$ be considered a simpler equivalent

equation to $\dfrac{2}{3}x + \dfrac{1}{5} = \dfrac{4}{7}$ ?

$$2x + \frac{3}{5} = \frac{12}{7}$$

$$5\left(2x + \frac{3}{5}\right) = 5 \cdot \frac{12}{7}$$

$$10x + 3 = \frac{60}{7}$$

Is $10x + 3 = \dfrac{60}{7}$ a simpler equivalent equation to $2x + \dfrac{3}{5} = \dfrac{12}{7}$ ?

$$10x + 3 = \frac{60}{7}$$

$$7(10x + 3) = 7 \cdot \frac{60}{7}$$

$$70x + 21 = 60$$

Is $70x + 21 = 60$ a simpler equivalent equation to $10x + 3 = \dfrac{60}{7}$ ?

$$70x + 21 = 60$$

$$70x = 39$$

$$x = \frac{39}{70}$$

Using the domain of rational numbers, the equation $\frac{2}{3}x + \frac{1}{5} = \frac{4}{7}$ has been

shown to be equivalent to $x = \frac{39}{70}$ and, therefore, has $\left\{\frac{39}{70}\right\}$ as its truth set.

Notice that we first developed an equivalent equation in which all the numbers were integers by simply multiplying by each denominator in turn until all the denominators were divided out.

This same procedure is used below on the next example, but notice the continued simplification of terms which shortens the process.

$$\frac{5}{6}x - \frac{7}{12} = \frac{3}{4}$$

$$6\left(\frac{5}{6}x - \frac{7}{12}\right) = 6 \cdot \frac{3}{4}$$

$$5x - \frac{7}{2} = \frac{9}{2}$$

$$2\left(5x - \frac{7}{2}\right) = 2 \cdot \frac{9}{2}$$

$$10x - 7 = 9$$

$$10x = 16$$

$$x = \frac{8}{5}$$

Using the domain of rational numbers, $\frac{5}{6}x - \frac{7}{12} = \frac{3}{4}$ is equivalent to

$10x - 7 = 9$ and has $\left\{\frac{8}{5}\right\}$ as its truth set.

Any equation involving rational numbers can be converted to an equivalent equation involving integers by multiplying by each denominator in turn and simplifying whenever possible.

EXERCISE 7.5.1

Using the domain of rational numbers find truth sets by first obtaining an equivalent equation involving only integers.

1.  $\frac{2}{5}x - \frac{1}{4} = \frac{1}{6}$
   
2.  $\frac{3}{4}x + \frac{2}{3} = \frac{5}{12}$

3. $\dfrac{2}{5}x + \dfrac{3}{4} = \dfrac{7}{10}$

4. $\dfrac{7}{8}x - \dfrac{5}{6} = \dfrac{1}{12}$

5. $\dfrac{3}{4}x - \dfrac{9}{10} = \dfrac{2}{5}$

6. $\dfrac{5}{3}x - \dfrac{1}{4} = \dfrac{7}{9}$

7. $\dfrac{5}{6}x - \dfrac{7}{12} = \dfrac{2}{3}$

8. $\dfrac{2}{7}x - \dfrac{3}{14} = \dfrac{1}{4}$

9. $\dfrac{2}{3}x + \dfrac{5}{12} = \dfrac{1}{4}$

10. $\dfrac{3}{8}x - \dfrac{1}{6} = \dfrac{5}{12}$

11. $\dfrac{1}{2}x + \dfrac{7}{10} = \dfrac{3}{5}$

12. $\dfrac{5}{3}x + \dfrac{2}{7} = \dfrac{1}{3}$

13. $\dfrac{1}{4}x + \dfrac{3}{8} = \dfrac{1}{2}$

14. $\dfrac{5}{12}x + \dfrac{2}{3} = \dfrac{1}{4}$

15. $\dfrac{2}{3}x - \dfrac{2}{5} = \dfrac{3}{5}x - \dfrac{1}{3}$

16. $\dfrac{5}{6}x - \dfrac{1}{3} = \dfrac{1}{4}x + \dfrac{1}{2}$

17. $\dfrac{3}{4}x - \dfrac{1}{2} = \dfrac{1}{4}x + \dfrac{11}{2}$

18. $\dfrac{1}{3}x + \dfrac{3}{4} = \dfrac{1}{2} - \dfrac{2}{3}x$

19. $\dfrac{2}{3}x + 1 = \dfrac{5}{9}x - \dfrac{1}{6}$

20. $\dfrac{5}{6}x - \dfrac{2}{9} = \dfrac{3}{4} - \dfrac{1}{2}x$

## 7.6  Equations with Variables in Denominators

In the preceding section a method for converting equations with rational numbers to equivalent equations with integers was explained. In this section that skill is expanded to include equations with the variable in the denominator.

One method of finding the truth set of $\dfrac{4}{3}x - \dfrac{5}{6} = \dfrac{3}{7}$ is to multiply both members of the equation by each denominator in turn until an equation with only integers is obtained. This would suggest that one method for solving $\dfrac{5}{x} - \dfrac{3}{2} = \dfrac{4}{x}$ would also include the successive multiplication of both members of the equation by the denominators, $x$ and 2. This method can and will be used, but it presents one different aspect that must be understood.

Is $\dfrac{5}{0}$ a rational number? No! Zero may never be used as the denomi-

nator of a rational number. Hence, $\dfrac{^-3}{0}, \dfrac{4}{0}$, and $\dfrac{0}{0}$ have no meaning as

rational numbers. What then about the truth or falsity of a statement like

$\dfrac{4}{0} + \dfrac{3}{2} = \dfrac{4}{0}$? The statement is certainly not true, but at this point perhaps

there is some doubt as to the origin of such a statement. Let us attempt

to solve $\dfrac{4}{x} - \dfrac{3}{2} = \dfrac{4}{x}$ remembering that any element of the truth set must

make $\dfrac{4}{x} - \dfrac{3}{2} = \dfrac{4}{x}$ a true statement.

(1)   First, multiply both members by the first denominator, $x$.

$$\dfrac{4}{x} - \dfrac{3}{2} = \dfrac{4}{x}$$

$$x\left(\dfrac{4}{x} - \dfrac{3}{2}\right) = x \cdot \dfrac{4}{x}$$

$$4 - \dfrac{3x}{2} = 4$$

(2)   Second, multiply both members by 2.

$$2\left(4 - \dfrac{3x}{2}\right) = 2 \cdot 4$$

$$8 - 3x = 8$$

(3)   Complete the solution.

$$^-3x = 0$$

$$x = 0$$

Zero is the only rational number which will make $x = 0$ a true statement,

but it should be clear that $\{0\}$ is not the truth set of $\dfrac{4}{x} - \dfrac{3}{2} = \dfrac{4}{x}$   because

$\dfrac{4}{0} - \dfrac{3}{2} = \dfrac{4}{0}$ is not a true statement. But what happened to our equation

solving method that brought us to the unacceptable result of $x = 0$?

The problem lies with the step in which both members of $\dfrac{4}{x} - \dfrac{3}{2} = \dfrac{4}{x}$

were multiplied by $x$. This step can not be justified by the Multiplicative Property of Equality because the property only guarantees equivalent equations when both members of the equation are multiplied by a nonzero number. The Multiplicative Property of Equality does not guarantee equivalent equations when an algebraic term, like $x$, is used as the multiplier.

Since zero is never allowed as a denominator, the first step in solving any fractional equation with a variable in the denominator is to exclude as possible elements of the truth sets those numbers which would result in zero denominators.

For the equation $\dfrac{3}{x} - \dfrac{4}{x + 2} = \dfrac{9}{x - 5}$ there are three numbers that

must be excluded from consideration as elements of the truth set. 0, ‾2, and 5 are not acceptable replacements for $x$ because they would lead to zero

denominators in the equation $\dfrac{3}{x} - \dfrac{4}{x + 2} = \dfrac{9}{x - 5}$ .

To exclude, as possible elements of the truth sets, those numbers which will cause zero denominators each of the denominators can be set equal to zero and solved for the variable. For example, to find excluded values for the

equation $\dfrac{6}{x - 3} + \dfrac{4}{x + 2} = \dfrac{{}^{-}5}{2x - 3}$ the following three equations are solved:

$$x - 3 = 0$$

$$x + 2 = 0$$

$$2x - 3 = 0$$

These equations were obtained by setting each of the denominators equal to zero. Since the truth sets of the three equations are $\{3\}$, $\{{}^{-}2\}$, and

$\left\{\dfrac{3}{2}\right\}$ , the elements of these sets are excluded as possible elements of the

truth set of $\dfrac{6}{x - 3} + \dfrac{4}{x + 2} = \dfrac{{}^{-}5}{2x - 3}$ .

EXERCISE 7.6.1

For each equation find those numbers which must be excluded as possible elements of the truth set because they would result in zero denominators.

1. $\dfrac{4}{x} - \dfrac{5}{x+1} = \dfrac{3}{x-2}$

6. $\dfrac{2}{x+8} - \dfrac{3}{2x} = \dfrac{4}{7}$

2. $\dfrac{2}{x+4} + \dfrac{4}{x-1} = \dfrac{1}{x+5}$

7. $\dfrac{5}{5x+1} - \dfrac{2}{2x-11} = \dfrac{7}{x-6}$

3. $\dfrac{1}{2x+1} = \dfrac{5}{x-4}$

8. $\dfrac{4}{x+9} + \dfrac{6}{5} = \dfrac{3}{5x-6}$

4. $\dfrac{3}{x+8} = \dfrac{2}{3x-8}$

9. $\dfrac{2}{x+1} = \dfrac{4}{x-7} + \dfrac{9}{11-2x}$

5. $\dfrac{7}{x-3} - \dfrac{2}{5-x} = \dfrac{7}{4+x}$

10. $\dfrac{7}{4x} - \dfrac{3}{5x} = \dfrac{9}{6-x}$

## 7.7  Solving Equations with Polynomial Fractions

From this point on in this chapter the domain of the variable is the set of rational numbers except in those instances where a zero denominator would be the result.

Recall that one method of solving $\dfrac{3}{4}x - \dfrac{5}{6} = \dfrac{1}{2}$ was the process of multiplying both sides of the equation by each denominator in turn, simplifying whenever possible, and obtaining an equation involving only integers.

$$\frac{3}{4}x - \frac{5}{6} = \frac{1}{2}$$

$$4\left(\frac{3}{4}x - \frac{5}{6}\right) = 4 \cdot \frac{1}{2}$$

$$3x - \frac{10}{3} = 2$$

$$3\left(3x - \frac{10}{3}\right) = 3 \cdot 2$$

$$9x - 10 = 6$$

$$9x = 16$$

$\left\{\dfrac{16}{9}\right\}$ is the truth set of $\dfrac{3}{4}x - \dfrac{5}{6} = \dfrac{1}{2}$.

To solve $\dfrac{5}{x-3} + 2 = \dfrac{2}{x-3}$ the first step is to multiply both members

by the denominator, $x - 3$. After simplifying, a relatively simple linear equation is solved for the truth set.

$$\frac{5}{x-3} + 2 = \frac{2}{x-3}$$

$$(x-3)\left(\frac{5}{x-3} + 2\right) = (x-3) \cdot \frac{2}{x-3}$$

$$5 + 2(x-3) = 2$$

$$5 + 2x - 6 = 2$$

$$2x - 1 = 2$$

$$2x = 3$$

$\left\{\dfrac{3}{2}\right\}$ is the truth set of $\dfrac{5}{x-3} + 2 = \dfrac{2}{x-3}$.

To solve $\dfrac{6}{2x+5} = \dfrac{4}{x-3}$ the equation will be multiplied by each

denominator in turn.

$$\frac{6}{2x+5} = \frac{4}{x-3}$$

$$(2x+5) \cdot \frac{6}{2x+5} = (2x+5) \cdot \frac{4}{x-3}$$

$$6 = \frac{4(2x+5)}{x-3}$$

$$(x-3) \cdot 6 = (x-3) \cdot \frac{4(2x+5)}{x-3}$$

$$6x - 18 = 8x + 20$$

$$^-2x = 38$$

$$x = {}^-19$$

$\{^-19\}$ is the truth set of $\dfrac{6}{2x+5} = \dfrac{4}{x-3}$.

EXERCISE 7.7.1

Find the truth set for each equation. At least one equation has { } as its truth set. Be certain to check possible truth set elements to avoid zero denominators.

1.  $\dfrac{5}{x} - 3 = \dfrac{2}{x}$

10. $\dfrac{6}{x - 5} = \dfrac{4}{x + 2}$

2.  $\dfrac{6}{x} + 9 = \dfrac{1}{x}$

11. $\dfrac{3}{x - 7} = \dfrac{5}{x - 1}$

3.  $\dfrac{5}{x} - 2 = \dfrac{11}{x}$

12. $\dfrac{6}{x + 5} = \dfrac{9}{x}$

4.  $\dfrac{8}{x - 1} + 7 = \dfrac{9}{x - 1}$

13. $\dfrac{3}{x - 7} = \dfrac{2}{x - 5}$

5.  $\dfrac{2}{x - 6} - 3 = \dfrac{1}{x - 6}$

14. $\dfrac{4}{x + 3} = \dfrac{2}{x - 3}$

6.  $\dfrac{5}{x + 4} + 2 = \dfrac{3}{x + 4}$

15. $\dfrac{5}{x + 1} = \dfrac{8}{x - 2}$

7.  $\dfrac{9}{x - 3} + 1 = \dfrac{9}{x - 3}$

16. $\dfrac{3}{x + 6} = \dfrac{5}{x + 10}$

8.  $\dfrac{4}{x + 2} + 6 = \dfrac{7}{x + 2}$

17. $\dfrac{6}{x - 7} = \dfrac{4}{x + 1}$

9.  $\dfrac{3}{x - 4} - 1 = \dfrac{1}{x - 4}$

18. $\dfrac{9}{x - 1} = \dfrac{7}{x - 3}$

19. $\dfrac{4}{x - 3} + \dfrac{2}{(x - 3)(x + 5)} = \dfrac{1}{x + 5}$

20. $\dfrac{2}{x + 8} + \dfrac{7}{(x + 8)(x + 1)} = \dfrac{5}{x + 1}$

The first step in solving equations involving polynomial fractions is to factor any denominators that are factorable. Once all denominators are factored, then the procedure is to multiply both members of the equation by each denominator in turn and simplify whenever possible.

To solve $\dfrac{3}{x+2} - \dfrac{6}{x^2 - 5x - 14} = \dfrac{5}{x-7}$ the following steps are used:

$$\frac{3}{x+2} - \frac{6}{x^2 - 5x - 14} = \frac{5}{x-7}$$

$$\frac{3}{x+2} - \frac{6}{(x+2)(x-7)} = \frac{5}{x-7}$$

$$(x+2)\left(\frac{3}{x+2} - \frac{6}{(x+2)(x-7)}\right) = (x+2) \cdot \frac{5}{x-7}$$

$$3 - \frac{6}{(x-7)} = \frac{5(x+2)}{x-7}$$

$$(x-7)\left(3 - \frac{6}{(x-7)}\right) = (x-7) \cdot \frac{5(x+2)}{x-7}$$

$$3(x-7) - 6 = 5(x+2)$$

$$3x - 21 - 6 = 5x + 10$$

$$3x - 27 = 5x + 10$$

$$^{-}2x = 37$$

$$x = \frac{^{-}37}{2}$$

$\dfrac{^{-}37}{2}$ is an acceptable replacement for the denominators $x+2$ and

$x - 7$ because it obviously does not produce zero. $\dfrac{^{-}37}{2}$ can be used as a

replacement for $x$ in the denominator $x^2 - 5x - 14$ to show that it also

does not lead to a zero denominator. Hence $\left\{\dfrac{^{-}37}{2}\right\}$ is the truth set of

$\dfrac{3}{x+2} - \dfrac{6}{x^2 - 5x - 14} = \dfrac{5}{x-7}$ .

Although it is suggested in the last paragraph that $\dfrac{-37}{2}$ be tried in the denominator $x^2 - 5x - 14$ to show that the result is not zero, the student should consider the following claim. The only replacements for $x$ that will cause $x^2 - 5x - 14$ to have an evaluation of zero are $^-2$ and $7$ because the factors of $x^2 - 5x - 14$ are $(x + 2)$ and $(x - 7)$. The concept involved in this claim will be important in the following section.

EXERCISE 7.7.2

Solve each equation. There is at least one equation with { } as its truth set. Check each possible element for the truth set to be certain that it does not lead to a zero denominator.

1. $\dfrac{4}{x - 5} - \dfrac{12}{x^2 - 7x + 10} = \dfrac{5}{x - 2}$

2. $\dfrac{9}{x + 3} - \dfrac{4}{x - 4} = \dfrac{2}{x^2 - x - 12}$

3. $\dfrac{2}{x + 1} + \dfrac{6}{x^2 + 3x + 2} = \dfrac{4}{x + 2}$

4. $\dfrac{7}{x - 2} - \dfrac{18}{x^2 + 2x - 8} = \dfrac{3}{x + 4}$

5. $\dfrac{5}{x + 5} - \dfrac{3}{x - 5} = \dfrac{2}{x^2 - 25}$

6. $\dfrac{2}{x - 1} - \dfrac{1}{x^2 - 1} = \dfrac{5}{x + 1}$

7. $\dfrac{15}{x^2 + 2x - 15} = \dfrac{4}{x - 3} - \dfrac{5}{x + 5}$

8. $\dfrac{7}{x + 3} - \dfrac{5}{x^2 + x - 6} = \dfrac{3}{x - 2}$

9. $\dfrac{5}{x - 7} - \dfrac{4}{x + 1} = \dfrac{3}{x^2 - 6x - 7}$

10.   $$\frac{8}{x+4} + \frac{5}{x^2+3x-4} = \frac{3}{x-1}$$

11.   $$\frac{8}{x^2+2x-8} = \frac{5}{x-2} - \frac{3}{x+4}$$

12.   $$\frac{5}{x+4} + \frac{12}{x^2+2x-8} = \frac{2}{x-2}$$

13.   $$\frac{1}{2x-1} - \frac{1}{2x^2+x-1} = \frac{1}{x+1}$$

14.   $$\frac{5}{x-3} + \frac{2}{x+4} = \frac{1}{x^2+x-12}$$

15.   $$\frac{9}{x+4} - \frac{1}{x^2-16} = \frac{3}{x-4}$$

16.   $$\frac{4}{x+7} = \frac{3}{x^2+8x+7} - \frac{5}{x+1}$$

17.   $$\frac{6}{x^2-81} = \frac{1}{x+9} + \frac{5}{x-9}$$

18.   $$\frac{4}{x-6} - \frac{5}{x^2-2x-24} = \frac{3}{x+4}$$

19.   $$\frac{3}{x+1} - \frac{7}{x^2-x-2} = \frac{5}{x-2}$$

20.   $$\frac{7}{x-6} + \frac{4}{x^2-36} = \frac{-2}{x+6}$$

## 7.8   The Use of the Multiplication Properties of Zero to Solve Equations

Zero is possessed of special number properties. In addition the sum of any number with zero is that same number. For this reason zero is called the identity element for addition.

In multiplication zero also has important properties. The first of these properties is that whenever zero is multiplied by any number then the product is zero. A second, and in this section most important, property of zero is that whenever the product of two or more numbers is zero then at least one of the numbers must be zero.

The truth set of $3x = 0$ is obvious because zero is the only number that can be multiplied by 3 to give zero.

In the equation $4(x - 2) = 0$ then 4 and $(x - 2)$ are multiplied together to give zero. Since at least one of the factors must itself be zero we can claim that $4 = 0$ or $x - 2 = 0$. $4 = 0$ is obviously false, but $x - 2 = 0$ leads to $\{2\}$ as the correct truth set for $4(x - 2) = 0$.

In the equation $5x(2x + 1)(x - 3) = 0$ there are three factors to be considered. Since the product of these three factors is zero, then at least one of the factors must itself be zero. This leads to the equations $5x = 0$,

$2x + 1 = 0$, $x - 3 = 0$ which have $\{0\}$, $\left\{\dfrac{-1}{2}\right\}$, and $\{3\}$ as their respective

truth sets. The truth set of $5x(2x + 1)(x - 3) = 0$ is $\left\{0, \dfrac{-1}{2}, 3\right\}$ because if

any one of the factors is zero then the product will be zero.

The equation $7(3x - 5)(x + 6) = 0$ is solved by considering the three equalities $7 = 0$, $3x - 5 = 0$, $x + 6 = 0$. Since $7 = 0$ is false, the equations $3x - 5 = 0$ and $x + 6 = 0$ will supply all the elements of the truth set.

Hence the truth set of $7(3x - 5)(x + 6) = 0$ is $\left\{\dfrac{5}{3}, {}^-6\right\}$ because $\left\{\dfrac{5}{3}\right\}$ is the

truth set of $3x - 5 = 0$ and $\{{}^-6\}$ is the truth set of $x + 6 = 0$.

EXERCISE 7.8.1

Find the truth set of each equation by recognizing that the only way a product of zero can be obtained is by having at least one factor equal to zero.

1. $4x(x - 3) = 0$
2. $2x(x + 8) = 0$
3. $(x - 3)(2x + 7) = 0$
4. $(x + 6)(3x - 1) = 0$
5. $(5x + 1)(2x + 3) = 0$
6. $x(4x - 1) = 0$
7. $3(5x + 3)(x + 9) = 0$
8. $x(x - 2)(x + 2) = 0$
9. $6x(x + 5)(x + 5) = 0$
10. $x(x - 4)(2x - 11) = 0$
11. $(x - 10)(x + 11) = 0$
12. $(4x + 7)(2x - 9) = 0$
13. $(7x + 3)(2x + 3) = 0$
14. $5(4x - 1)(x - 8) = 0$
15. $3x(x + 2)(x + 2) = 0$
16. $7(3x - 14)(x + 1) = 0$
17. $x^2(5x + 7) = 0$
18. $2(x - 3)(x - 3) = 0$
19. $(5 - x)(7 + x) = 0$
20. $6(19 + 3x)(12 - x) = 0$

## 7.9  Solving Quadratic Equations with Rational Solutions

Equations such as $4x - 3 = 7$ and $\dfrac{3}{4}x - \dfrac{5}{6} = \dfrac{2}{3}x + \dfrac{1}{7}$ are linear equations. The variable in each case has an implied exponent of 1.

Equations such as $x^2 - 5x - 6 = 0$ and $x^2 + 2x = 8$ are called quadratic equations. The variable in each case has an exponent of 2 in one of its terms.

In this section the solution of quadratic equations with rational number solutions is studied. The solution of such equations is heavily dependent upon the concepts of the last exercise. For the equation $4x(x - 3)(2x + 1) = 0$ the three following linear equations are solved.

$$4x = 0 \qquad \{0\}$$

$$x - 3 = 0 \qquad \{3\}$$

$$2x + 1 = 0 \qquad \left\{\dfrac{-1}{2}\right\}$$

Hence the truth set of $4x(x - 3)(2x + 1) = 0$ is $\left\{0, 3, \dfrac{-1}{2}\right\}$.

To solve $x^2 - 5x - 6 = 0$ the first step is to factor the trinomial $x^2 - 5x - 6$. Then the procedure is identical to the solutions of the last exercise.

$$x^2 - 5x - 6 = 0$$

$$(x - 6)(x + 1) = 0$$

$$x - 6 = 0 \quad \text{or} \quad x + 1 = 0$$

Hence the truth set of $x^2 - 5x - 6 = 0$ is $\{6, {}^-1\}$.

The polynomial $x^2 + 17x + 6$ is prime; it can not be factored. Since $x^2 + 17x + 6$ is prime the equation $x^2 + 17x + 6 = 0$ has no rational number solutions and the truth set is $\{\ \ \}$ when the domain of rational numbers is used.

$2x^2 - 11x$ is factored as $x(2x - 11)$ and $2x^2 - 11x = 0$ is solved as shown below:

$$2x^2 - 11x = 0$$

$$x(2x - 11) = 0$$

$$x = 0 \quad \text{or} \quad 2x - 11 = 0$$

Hence $\left\{0, \dfrac{11}{2}\right\}$ is the truth set of $2x^2 - 11x = 0$

EXERCISE 7.9.1

Find the truth set using the domain of rational numbers. Completely factor each polynomial.

| | | | |
|---|---|---|---|
| 1. | $x^2 + 7x + 12 = 0$ | 16. | $x^2 + 8x + 7 = 0$ |
| 2. | $x^2 - 5x - 14 = 0$ | 17. | $x^2 + 3x + 2 = 0$ |
| 3. | $x^2 - 9 = 0$ | 18. | $x^2 - 10x + 21 = 0$ |
| 4. | $x^2 - 8x = 0$ | 19. | $x^2 + 11x + 16 = 0$ |
| 5. | $x^2 - 11x + 10 = 0$ | 20. | $3x^2 + 6x - 45 = 0$ |
| 6. | $2x^2 - 7x + 5 = 0$ | 21. | $7x^2 - 7 = 0$ |
| 7. | $x^2 - 49 = 0$ | 22. | $4x^2 - 12x - 40 = 0$ |
| 8. | $x^2 + 6x + 8 = 0$ | 23. | $6x^2 - 7x - 10 = 0$ |
| 9. | $x^2 + 3x = 0$ | 24. | $3x^2 - 16x + 5 = 0$ |
| 10. | $x^2 - 4 = 0$ | 25. | $3x^2 - 8x - 3 = 0$ |
| 11. | $x^2 + 6x + 9 = 0$ | 26. | $4x^2 - 1 = 0$ |
| 12. | $x^2 + 5x - 24 = 0$ | 27. | $x^2 + 20x + 100 = 0$ |
| 13. | $-3x^2 + 12x = 0$ | 28. | $9x^2 - 4 = 0$ |
| 14. | $x^2 + 5x + 8 = 0$ | 29. | $5x^2 - 8x - 4 = 0$ |
| 15. | $x^2 - 5x + 4 = 0$ | 30. | $2x^2 - 7x + 6 = 0$ |

The truth sets of the preceding exercise were completely dependent upon the multiplication properties of zero that claim that zero times any number is zero and whenever the product of two numbers is zero then at least one of the factors must also be zero. Zero is the only rational number that has these special multiplication properties.

To solve the equation $x^2 - 2x - 3 = 5$ it would be of no value whatsoever to factor $x^2 - 2x - 3$ because the right member of the equation is 5. Zero is needed as the product of the factors. Consequently, $x^2 - 2x - 3 = 5$ is solved in the following steps:

$$x^2 - 2x - 3 = 5$$

$$(x^2 - 2x - 3) - 5 = 5 - 5$$

$$x^2 - 2x - 8 = 0$$

$$(x - 4)(x + 2) = 0$$

$$x - 4 = 0 \quad \text{or} \quad x + 2 = 0$$

$\{4, {}^-2\}$ is the truth set of $x^2 - 2x - 3 = 5$.

There are three elements in the truth set of $2x^3 - 7x^2 = 4x$. The truth set is found as shown below.

$$2x^3 - 7x^2 = 4x$$

$$(2x^3 - 7x^2) - 4x = 4x - 4x$$

$$2x^3 - 7x^2 - 4x = 0$$

$$x(2x^2 - 7x - 4) = 0$$

$$x(x - 4)(2x + 1) = 0$$

$$x = 0 \quad \text{or} \quad x - 4 = 0 \quad \text{or} \quad 2x + 1 = 0$$

The truth set of $2x^3 - 7x^2 = 4x$ is $\left\{0, 4, \dfrac{-1}{2}\right\}$.

EXERCISE 7.9.2

Find the truth set using the domain of rational numbers.

| | | | |
|---|---|---|---|
| 1. | $x^2 - 8x = {}^-15$ | 11. | $4x^2 = 9$ |
| 2. | $x^2 = 17x$ | 12. | $12x - 36 = x^2$ |
| 3. | $x^2 + 3x = 10$ | 13. | $x^2 + 4x = {}^-4$ |
| 4. | $x^2 + 12 = 7x$ | 14. | $x^2 - 72 = x$ |
| 5. | $x^2 = 9x - 8$ | 15. | $3x + 28 = x^2$ |
| 6. | $x^2 + x = 30$ | 16. | $x^2 + 17x = 18$ |
| 7. | $2x^2 - 5x = 12$ | 17. | $x^3 - 5x = 4x^2$ |
| 8. | $12x^2 = 15x$ | 18. | $x^2 = 9x + 10$ |
| 9. | $x^2 + 9 = 6x$ | 19. | $x^2 + 25 = 10x$ |
| 10. | $x^2 = 25$ | 20. | $2x - x^2 = x^3$ |

# 8 THE REAL NUMBERS

## 8.1 Introduction

In chapters 2, 3, and 4 three different sets of numbers were studied along with their properties. Chapter 2 was a presentation of the system of counting numbers, $\{1, 2, 3, \ldots\}$. In chapter 3 we developed the system of integers, $\{\ldots, {}^-2, {}^-1, 0, 1, 2, \ldots\}$. Then in chapter 4, we studied the system of rational numbers which depends on the set of all numbers of the form $\frac{x}{y}$ where $x$ and $y$ are integers and $y \neq 0$.

Each of the number systems presented preserved the properties enjoyed by any forerunners and also included new properties which proved useful in finding nonempty truth sets for linear equations. In fact, the system of rational numbers is capable of supplying a nonempty truth set for every linear equation of the form $ax + b = c$ where $a$, $b$, and $c$ are rational numbers and $a \neq 0$.

In this chapter we shall further expand the set of rational numbers to include all numbers which are called irrational. The union of the set of rational numbers with the set of irrational numbers is called the set of real numbers.

## 8.2    The Density of the Rational Numbers

We say that the rational numbers are *dense* because whenever two distinct rational numbers are chosen then there is always some rational numbers between them.

For example, suppose the rational numbers 0 and 7 were chosen and you were asked to find a set of ten rational numbers between them. Certainly 1, 2, 3, 4, 5, and 6 are relatively easy choices to make, but we are not limited to integers so we can also include rational numbers such as $\frac{1}{2}, \frac{5}{6}, \frac{17}{8}$, and $\frac{47}{13}$. One possible correct answer for the problem could be

$$\left\{1, 2, 3, 4, 5, 6, \frac{1}{2}, \frac{5}{6}, \frac{17}{8}, \frac{47}{13}\right\}.$$

Suppose that the problem were to find ten rational numbers between $\frac{-13}{5}$ and $\frac{-2}{5}$. One possible set would be $\left\{\frac{-12}{5}, \frac{-11}{5}, \frac{-10}{5}, \frac{-9}{5}, \frac{-8}{5}, \frac{-7}{5}, \frac{-6}{5}, \frac{-5}{5}, \frac{-4}{5}, \frac{-3}{5}\right\}$.

Do you see how the elements of the set were selected?

The preceding two examples offered fairly simple solutions because of the rational numbers chosen in stating the problems. This would not be the case if you were asked to find ten rational numbers between $\frac{5}{6}$ and $\frac{13}{17}$. One method of finding the set could be accomplished by first writing $\frac{5}{6}$ and $\frac{13}{17}$ using the same denominator. Since $6 \cdot 17$ is 102 we can find that $\frac{5}{6} = \frac{85}{102}$ and $\frac{13}{17} = \frac{78}{102}$. Hence the problem could be rephrased as requesting ten rational numbers between $\frac{85}{102}$ and $\frac{78}{102}$. Now, six rational numbers between $\frac{5}{6}\left(\frac{85}{102}\right)$ and $\frac{13}{17}\left(\frac{78}{102}\right)$ should seem fairly obvious if integers between 78 and 85 are chosen as numerators for a denominator of 102.

$\dfrac{79}{102}, \dfrac{80}{102}, \dfrac{81}{102}, \dfrac{82}{102}, \dfrac{83}{102}, \dfrac{84}{102}$ are six rational numbers between $\dfrac{5}{6}$ and $\dfrac{13}{17}$,

but since ten numbers were requested the denominators can be changed

once again. By doubling 102, we find that $\dfrac{5}{6} = \dfrac{85}{102} = \dfrac{170}{204}$ and $\dfrac{13}{17} = \dfrac{78}{102} = \dfrac{156}{204}$ .

Finally we find ten rational numbers between $\dfrac{5}{6}$ and $\dfrac{13}{17}$ by using $\dfrac{170}{204}$ and

$\dfrac{156}{204}$ . One correct set would be $\left\{ \dfrac{157}{204}, \dfrac{158}{204}, \dfrac{159}{204}, \dots, \dfrac{166}{204} \right\}$ .

EXERCISE 8.2.1

For each pair of rational numbers find a set of ten rational numbers such that each of the ten is between the two given numbers.

1. 9 and 20

6. $\dfrac{-11}{12}$ and $\dfrac{-17}{12}$

2. $^-8$ and 0

7. $\dfrac{4}{7}$ and $\dfrac{2}{3}$

3. 5 and 7

8. $\dfrac{1}{2}$ and $\dfrac{9}{20}$

4. 2 and 3

9. $\dfrac{3}{5}$ and $\dfrac{2}{3}$

5. $\dfrac{4}{5}$ and $\dfrac{7}{5}$

10. $\dfrac{7}{9}$ and $\dfrac{13}{19}$

The procedure used in the last exercise could be broadened to allow us to find a hundred, or a thousand, or even a million rational numbers between any two distinct rational numbers. The arithmetic would become rather lengthy and cumbersome, but the thinking remains much the same.

Hence we say the rational numbers are dense. They are clustered together about every point on the number line. If we were to begin naming all the rational numbers between 1 and 2 we would devote the rest of our lives to the task. The job can not be finished because it has no end. When we say the rational numbers are dense we mean that there is no limit to the rational numbers between any two distinct points on the number line.

This discussion of the density of the rational numbers often leads to the intuitive conclusion that the rational numbers would "fill up" the number line if the attempt were made to assign every rational number to its associated point on the line. It is surprising, therefore, that if the number line is considered as a continuous string of points there are many such points that can never be associated with any rational numbers. Such points are associated with numbers which are called irrational.

### 8.3  System of Real Numbers

If the set of irrational numbers is unioned with the set of rational numbers then the result is a set of numbers which is capable of naming every point on the number line. This set of numbers which contains all the rationals and irrationals is called the set of real numbers.

The system of real numbers is a mathematical field and satisfies the following properties:

$C_1$:  The sum of two real numbers is a real number.

$C_2$:  The product of two real numbers is a real number.

The following open sentences become true statements for all real number replacements of $x$, $y$, and $z$.

$A_1$:  Commutative Law of Addition

$$x + y = y + x$$

$A_2$:  Associative Law of Addition

$$(x + y) + z = x + (y + z)$$

$A_3$:  Zero Is the Identity Element for Addition

$$x + 0 = x$$

$A_4$:  Every Real Number Has an Opposite. For any real number $x$ there is a real number $y$ such that $x + y = 0$.

$M_1$:  Commutative Law of Multiplication

$$xy = yx$$

$M_2$:  Associative Law of Multiplication

$$(xy)z = x(yz)$$

$M_3$:  One Is the Identity Element for Multiplication

$$1x = x$$

$M_4$:  Every Nonzero Real Number Has a Reciprocal. For any nonzero real number $x$ there is a real number $y$ such that $xy = 1$.

$D_1$:  The Distributive Law of Multiplication over Addition

$$x(y + z) = xy + xz$$

$O_1$:  Trichotomy Order Property. Exactly one of the following is true:

$$x = y \qquad x > y \qquad x < y$$

$O_2$:  The Addition Property of Inequalities. If $x > y$ then $x + z > y + z$.

$O_3$:  The Multiplication Property of Inequalities. If $x > y$ and $z > 0$ then $xz > yz$ and if $x > y$ and $z < 0$ then $xz < yz$.

*The Completeness Property*

There is a one-to-one correspondence between the set of real numbers and the set of all points on the number line.

EXERCISE 8.3.1

$\sqrt{2}$, $\pi$, and $\sqrt[3]{30}$ are three numerals for irrational numbers. In this exercise, each problem is a statement. Cite the property of the real numbers which states that the statement is true.

1.  $\sqrt{2} \cdot 7 = 7 \cdot \sqrt{2}$
2.  $1 \cdot \sqrt[3]{30} = \sqrt[3]{30}$
3.  Since $\sqrt{2} < \sqrt[3]{30}$ then $\sqrt{2} + \pi < \sqrt[3]{30} + \pi$
4.  $(\sqrt[3]{30} + \pi) + \sqrt{2} = \sqrt[3]{30} + (\pi + \sqrt{2})$
5.  Exactly one of the following is true:

$$\sqrt[3]{30} = \pi \qquad \sqrt[3]{30} < \pi \qquad \sqrt[3]{30} > \pi$$

6.  Since $\pi$ is a real number there is another real number that can be added to it to give a sum of zero.
7.  $\sqrt[3]{30} + \sqrt{2} = \sqrt{2} + \sqrt[3]{30}$
8.  $\sqrt{2} \cdot \pi + \sqrt{2} \cdot \sqrt[3]{30} = \sqrt{2}(\pi + \sqrt[3]{30})$
9.  Since $\sqrt{2}$ is a nonzero real number there is another real number that can be multiplied by it to give a product of one.
10.  $(\sqrt{2} \cdot \sqrt[3]{30}) \cdot \pi = \sqrt{2}(\sqrt[3]{30} \cdot \pi)$

## 8.4  Perfect Square Integers

When we speak of "squaring" an integer we mean to multiply that number by itself. Another method for showing that a number is to be multiplied by itself is accomplished by the use of an exponent of 2. Hence, the directions of "squaring 7" will be completed by evaluating $7^2$.

The numbers that are obtained by squaring integers are called *perfect square integers*. Therefore, since the evaluation of $7^2$ is 49 we say that 49 is a *perfect square integer*. Similarly, since the evaluation of $(^-4)^2$ is 16, we say that 16 is a perfect square integer.

Integers such as 5, 31, $^-9$, and 41 are not perfect square integers because in each case there is no integer that can be squared to give such products. Integers such as 4, 25, 100, and 625 are perfect square integers because in each case there is at least one integer that can be squared to give these integers as products.

$$2^2 = 4$$

$$(^-5)^2 = 25$$

$$(^-10)^2 = 100$$

$$25^2 = 625$$

EXERCISE 8.4.1

1.  Evaluate:
    (a)  $(^-2)^2$                    (d)  $(^-25)^2$
    (b)  $5^2$                        (e)  $(^-6)^2$
    (c)  $10^2$

2.  Each of the following numbers is a perfect square integer. Find a *counting number* that can be squared to give it as the product.
    (a)  9                            (d)  324
    (b)  64                           (e)  676
    (c)  169

3.  Each of the following numbers is a perfect square integer. Find a *negative integer* that can be squared to give it as the product.
    (a)  1                            (d)  196
    (b)  81                           (e)  441
    (c)  144

4.  Is zero a perfect square integer? Why?

5. Are there any negative numbers that are perfect square integers? Why?

6. Between each pair of numbers given below there are two perfect square integers. Find them.

   (a) 6 and 23        (d) 950 and 1050
   (b) 40 and 75       (e) ⁻53 and 3
   (c) 325 and 405

## 8.5  The Radical Sign

"$\sqrt{\phantom{x}}$" is a mathematical symbol called the radical sign. When the radical sign is placed over a real number numeral, that real number is called the radicand. In the expression "$\sqrt{17}$" the radical sign is used with the real number numeral "17" and 17 is called the radicand. Similarly $\sqrt{47}$, $\sqrt{64}$, and $\sqrt{⁻9}$ are three examples of the use of the radical sign where the radicands are 47, 64, and ⁻9 respectively.

"$\sqrt{4}$" is a numeral which uses the radical sign and has 4 as its radicand. "$\sqrt{4}$" is a numeral for 2 because the use of the radical sign names the *non-negative* real number that when squared would give 4.

"$\sqrt{25}$" is a numeral for 5. The radical sign indicates that $\sqrt{25}$ is the *non-negative* real number which when squared would give 25, the radicand.

Although both 5 and ⁻5 can be squared to give 25, the use of the radical sign in $\sqrt{25}$ requires that the non-negative number be used. Hence $\sqrt{25} = 5$ and $\sqrt{25} \neq ⁻5$.

Whenever the symbol "$\sqrt{\phantom{x}}$" is used it is read as "the principal square root of" its radicand. Hence, "$\sqrt{25}$" is read as "the principal square root of 25." The principal square root of a number can not be negative.

The symbol "$⁻\sqrt{\phantom{x}}$" is read as "the opposite of the principal square root" of its radicand. "$⁻\sqrt{25}$" is read as "the opposite of the principal square root of 25." Hence, $⁻\sqrt{25} = ⁻5$ and $⁻\sqrt{25} \neq 5$. The opposite of the principal square root of a number can never be positive.

EXERCISE 8.5.1

Find the integer named by each of the following numerals.

1. $\sqrt{49}$          7. $⁻\sqrt{289}$
2. $\sqrt{9}$           8. $\sqrt{784}$
3. $⁻\sqrt{36}$         9. $⁻\sqrt{121}$
4. $⁻\sqrt{81}$        10. $⁻\sqrt{0}$
5. $\sqrt{0}$          11. $\sqrt{25}$
6. $\sqrt{225}$        12. $\sqrt{196}$

13. $\sqrt{400}$          17. $\sqrt{169}$

14. $^-\sqrt{1}$          18. $\sqrt{64}$

15. $\sqrt{1024}$          19. $^-\sqrt{4}$

16. $\sqrt{16}$          20. $^-\sqrt{900}$

$\sqrt{49}$ is a numeral for 7 because it names the non-negative real number that when squared will give 49.

$\sqrt{50}$ is a numeral for a real number, but that real number is not an integer because no integer can be squared to give 50. Although $\sqrt{50}$ is not an integer, its size can be estimated. Since $\sqrt{49} = 7$ then it should be obvious that $\sqrt{50} > 7$. Also, since $\sqrt{64} = 8$ then $8 > \sqrt{50}$. Therefore, $\sqrt{50}$ must be between 7 and 8 because 50 is between the perfect square integers 49 and 64.

$^-\sqrt{20}$ is not an integer because there is no negative integer that can be squared to give 20. Nevertheless, $^-\sqrt{20}$ can be estimated as between $^-5$ and $^-4$ simply because 20 is between the perfect square integers 16 and 25.

The placement of the opposite dash is extremely important when using the radical sign. When the opposite dash is outside the radical sign then it generally poses no difficulty. $^-\sqrt{81}$ names the integer $^-9$ and $^-\sqrt{30}$ names a real number that is between $^-6$ and $^-5$. When the opposite dash applies to the radicand and makes it negative then the radical numeral does not name any real number whatsoever. $\sqrt{^-16}$ does not name a real number because the radicand is negative. The square of any real number can not be negative and, therefore, numerals such as $\sqrt{^-16}$, $\sqrt{^-5}$, $\sqrt{^-105}$, and $\sqrt{^-41}$ do not name real numbers.

EXERCISE 8.5.2

Each problem has a radical sign numeral. If it names an integer then find it. If it names another real number then name the two integers it is between. If it does not name a real number then state that is the case.

1. $\sqrt{12}$          11. $\sqrt{96}$

2. $^-\sqrt{9}$          12. $\sqrt{150}$

3. $\sqrt{^-25}$          13. $\sqrt{^-47}$

4. $^-\sqrt{69}$          14. $\sqrt{2}$

5. $\sqrt{100}$          15. $^-\sqrt{3}$

6. $\sqrt{8}$          16. $\sqrt{36}$

7. $^-\sqrt{43}$          17. $^-\sqrt{180}$

8. $^-\sqrt{110}$          18. $\sqrt{5}$

9. $^-\sqrt{^-7}$          19. $\sqrt{^-80}$

10. $\sqrt{64}$          20. $^-\sqrt{39}$

### 8.6 Perfect Square Rational Numbers

$\dfrac{25}{4}$ is called a perfect square rational number because there is a

rational number that can be squared to give $\dfrac{25}{4}$. $\left(\dfrac{5}{2}\right)^2 = \dfrac{25}{4}$ and also

$\left(\dfrac{-5}{2}\right)^2 = \dfrac{25}{4}$.

$\dfrac{17}{36}$ is not a perfect square rational number because there is no rational

number that can be squared to give $\dfrac{17}{36}$.

At first glance $\dfrac{18}{8}$ may not seem to be a perfect square rational number.

It is, but $\dfrac{18}{8}$ should be simplified to $\dfrac{9}{4}$ to make it more apparent that

$\left(\dfrac{3}{2}\right)^2 = \dfrac{9}{4}$ and $\left(\dfrac{-3}{2}\right)^2 = \dfrac{9}{4}$.

The determination of perfect square rational numbers is accomplished in two steps:

(1)   Always simplify the rational number.

(2)   Whenever both the numerator and denominator are perfect square integers then the rational number is also a perfect square.

The radical sign is used with rational number radicands in the same

manner as it was with integers as radicands. "$\sqrt{\dfrac{25}{4}}$" is read as "the prin-

cipal square root of $\dfrac{25}{4}$." The principal square root is never negative so

$\sqrt{\dfrac{25}{4}} = \dfrac{5}{2}$ and $\sqrt{\dfrac{25}{4}} \neq \dfrac{-5}{2}$.

" $^{-}\sqrt{\dfrac{25}{4}}$ " is read as "the opposite of the principal square root of $\dfrac{25}{4}$ ."

Hence $^{-}\sqrt{\dfrac{25}{4}} = \dfrac{-5}{2}$ and $^{-}\sqrt{\dfrac{25}{4}} \neq \dfrac{5}{2}$ .

As before any numeral with a negative radicand can not name a real number because the square of any real number gives a non-negative product.

EXERCISE 8.6.1

For each radical numeral below name the rational number it represents or state that it does not represent any real number.

1. $\sqrt{\dfrac{9}{49}}$     11. $\sqrt{\dfrac{49}{9}}$

2. $^{-}\sqrt{\dfrac{25}{64}}$     12. $\sqrt{\dfrac{81}{100}}$

3. $^{-}\sqrt{\dfrac{1}{4}}$     13. $^{-}\sqrt{\dfrac{25}{64}}$

4. $\sqrt{\dfrac{2}{8}}$     14. $\sqrt{\dfrac{-9}{16}}$

5. $\sqrt{\dfrac{12}{27}}$     15. $^{-}\sqrt{\dfrac{144}{25}}$

6. $\sqrt{\dfrac{-81}{4}}$     16. $\sqrt{\dfrac{4}{121}}$

7. $^{-}\sqrt{\dfrac{100}{9}}$     17. $^{-}\sqrt{\dfrac{5}{45}}$

8. $\sqrt{\dfrac{32}{98}}$     18. $\sqrt{\dfrac{-36}{81}}$

9. $\sqrt{\dfrac{-25}{16}}$     19. $^{-}\sqrt{\dfrac{49}{400}}$

10. $^{-}\sqrt{\dfrac{4}{49}}$     20. $\sqrt{\dfrac{900}{121}}$

## 8.7  Multiplying Principal Square Roots

The multiplication of a rational number and a principal square root is indicated by placing the numbers in juxtaposition to each other. $7\sqrt{3}$ means the same as $7 \cdot \sqrt{3}$.  $\dfrac{-3}{4}\sqrt{6}$ means the same as $\dfrac{-3}{4} \cdot \sqrt{6}$.

The opposite of the square root of 13 is written as $^-\sqrt{13}$, but also can be written as $^-1\sqrt{13}$ or $^-1 \cdot \sqrt{13}$. Similarly $^-\sqrt{83}$ is $^-1\sqrt{83}$ or $^-1 \cdot \sqrt{83}$ and $^-\sqrt{47}$ is $^-1\sqrt{47}$ or $^-1 \cdot \sqrt{47}$.

The product of two principal square roots is accomplished by multiplying their radicands. To multiply $\sqrt{2}$ and $\sqrt{3}$ we multiply 2 and 3 under a common radical sign and use their product, 6, as the new radicand.

$$\sqrt{2} \cdot \sqrt{3} = \sqrt{2 \cdot 3} = \sqrt{6}$$

Four more examples of multiplication of principal square roots are shown below. In each case, the example shows that the product of two principal square roots is the principal square root of their radicands.[1]

$$\sqrt{6} \cdot \sqrt{7} = \sqrt{6 \cdot 7} = \sqrt{42}$$

$$\sqrt{5} \cdot \sqrt{11} = \sqrt{5 \cdot 11} = \sqrt{55}$$

$$\sqrt{2} \cdot \sqrt{13} = \sqrt{2 \cdot 13} = \sqrt{26}$$

$$\sqrt{5} \cdot \sqrt{7} = \sqrt{5 \cdot 7} = \sqrt{35}$$

Because multiplication of real numbers satisfies both the commutative and associative properties, we can now explain the multiplication of problems such as $7\sqrt{3} \cdot {}^-6\sqrt{2}$ and $\sqrt{5} \cdot 3\sqrt{11}$.

The commutative and associative properties of multiplication allow the order and grouping of any factors to be changed. In the problem $7\sqrt{3} \cdot {}^-6\sqrt{2}$, four factors can be considered. They are 7, $\sqrt{3}$, $^-6$, and $\sqrt{2}$. Changing the order and grouping of these factors results in the following multiplication.

$$7\sqrt{3} \cdot {}^-6\sqrt{2} = (7 \cdot {}^-6) \cdot (\sqrt{3} \cdot \sqrt{2})$$

$$= {}^-42 \cdot \sqrt{6}$$

$$= {}^-42\sqrt{6}$$

[1] The method of multiplication described here applies only to real numbers. It should not be considered appropriate for those radicals with negative radicands.

The problem $\sqrt{5} \cdot 3\sqrt{11}$ is completed in the following steps.

$$\sqrt{5} \cdot 3\sqrt{11} = 3 \cdot (\sqrt{5} \cdot \sqrt{11})$$

$$= 3 \cdot \sqrt{55}$$

$$= 3\sqrt{55}$$

EXERCISE 8.7.1

1.  Multiply:

    (a) $\sqrt{3} \cdot \sqrt{5}$        (j) $2\sqrt{5} \cdot {}^-\sqrt{6}$

    (b) $\sqrt{2} \cdot \sqrt{7}$        (k) ${}^-3\sqrt{7} \cdot {}^-2\sqrt{13}$

    (c) $\sqrt{6} \cdot \sqrt{11}$        (l) $5\sqrt{3} \cdot {}^-\sqrt{10}$

    (d) $\sqrt{5} \cdot \sqrt{17}$        (m) ${}^-2\sqrt{19} \cdot 5\sqrt{2}$

    (e) $\sqrt{10} \cdot \sqrt{3}$        (n) $8\sqrt{2} \cdot \sqrt{13}$

    (f) $\sqrt{19} \cdot \sqrt{2}$        (o) ${}^-\sqrt{5} \cdot {}^-\sqrt{14}$

    (g) $\sqrt{11} \cdot \sqrt{10}$        (p) $2 \cdot {}^-3\sqrt{19}$

    (h) $\sqrt{13} \cdot \sqrt{15}$        (q) $3\sqrt{2} \cdot \sqrt{5}$

    (i) $4\sqrt{17} \cdot 5\sqrt{2}$        (r) ${}^-4 \cdot {}^-\sqrt{6}$

2.  Find the missing factor in each problem.

    (a) $\sqrt{4} \cdot \underline{\quad} = \sqrt{12}$

    (b) $\sqrt{9} \cdot \underline{\quad} = \sqrt{54}$

    (c) $\sqrt{25} \cdot \underline{\quad} = \sqrt{125}$

    (d) $\sqrt{36} \cdot \underline{\quad} = \sqrt{72}$

    (e) $\sqrt{16} \cdot \underline{\quad} = \sqrt{48}$

    (f) $\sqrt{100} \cdot \underline{\quad} = \sqrt{200}$

    (g) $\sqrt{64} \cdot \underline{\quad} = \sqrt{320}$

    (h) $\sqrt{49} \cdot \underline{\quad} = \sqrt{294}$

    (i) $\sqrt{81} \cdot \underline{\quad} = \sqrt{810}$

    (j) $\sqrt{900} \cdot \underline{\quad} = \sqrt{4500}$

## 8.8  Simplifying Radical Expressions

Expressions such as $\sqrt{36}$ and $\sqrt{\dfrac{25}{49}}$ name rational numbers because in each case the radicand is a perfect square.

If the radicand of a radical expression is positive, but not a perfect square, then the number it names is an irrational number. For example, $\sqrt{12}$ is a radical expression which names an irrational number because 12 is a positive number that is not a perfect square.

In this section we are interested in simplifying radical expressions with the emphasis on those which name irrational numbers. Nevertheless, the procedure for simplifying radical expressions is dependent upon a knowledge of the perfect square integers, 0, 1, 4, 9, 16, 25, 36, 49, 64, 81, 100, 121, 144, etc.

To simplify $\sqrt{12}$ the first step is to look for a perfect square integer that is a factor of 12. In this case 4 is the greatest perfect square integer that is a factor of 12, so $\sqrt{12}$ is written as a multiplication problem where one factor is $\sqrt{4}$. This forces the other factor to be $\sqrt{3}$ and the simplification of $\sqrt{12}$ is done as shown below.

$$\sqrt{12} = \sqrt{4} \cdot \sqrt{3}$$
$$= 2 \cdot \sqrt{3}$$
$$= 2\sqrt{3}$$

To simplify $3\sqrt{72}$, the first step is to find a perfect square integer that is a factor of 72. In this case 4, 9, and 36 are all factors of 72, but the factor 36 is greatest and it should be used to avoid the necessity for further simplification.

$$3\sqrt{72} = 3\sqrt{36} \cdot \sqrt{2}$$
$$= 3 \cdot 6 \cdot \sqrt{2}$$
$$= 18\sqrt{2}$$

Not all radical expressions can be simplified. Such an expression is $\sqrt{10}$. The only perfect square factor of 10 is 1 and it accomplishes little, if anything, to write $\sqrt{10}$ as $\sqrt{1} \cdot \sqrt{10}$.

EXERCISE 8.8.1

Simplify:

| | | | |
|---|---|---|---|
| 1. | $\sqrt{50}$ | 12. | $^{-}\sqrt{108}$ |
| 2. | $\sqrt{8}$ | 13. | $3\sqrt{175}$ |
| 3. | $\sqrt{18}$ | 14. | $5\sqrt{90}$ |
| 4. | $\sqrt{75}$ | 15. | $^{-}\sqrt{300}$ |
| 5. | $\sqrt{32}$ | 16. | $2\sqrt{162}$ |
| 6. | $\sqrt{54}$ | 17. | $^{-7}\sqrt{20}$ |
| 7. | $\sqrt{24}$ | 18. | $7\sqrt{63}$ |
| 8. | $\sqrt{48}$ | 19. | $5\sqrt{1100}$ |
| 9. | $\sqrt{150}$ | 20. | $^{-3}\sqrt{49}$ |
| 10. | $\sqrt{60}$ | 21. | $5\sqrt{45}$ |
| 11. | $2\sqrt{27}$ | 22. | $6\sqrt{52}$ |

23. $-3\sqrt{96}$          27. $-6\sqrt{225}$
24. $2\sqrt{288}$          28. $8\sqrt{92}$
25. $-4\sqrt{500}$         29. $11\sqrt{250}$
26. $3\sqrt{99}$           30. $-7\sqrt{256}$

## 8.9  Multiplying Fractions with Radical Expressions

One of the field properties that applies to the system of real numbers states that every nonzero real number has a reciprocal. $\sqrt{2}$ is a nonzero real number so the property assures us it has a reciprocal. The reciprocal of

$\sqrt{2}$ is often designated by the numeral $\dfrac{1}{\sqrt{2}}$. Similarly, $\sqrt{7}$ has a reciprocal

designated by $\dfrac{1}{\sqrt{7}}$, $2\sqrt{21}$ has a reciprocal designated by $\dfrac{1}{2\sqrt{21}}$, and

$-\sqrt{13}$ has a reciprocal designated by $\dfrac{1}{-\sqrt{13}}$ or $\dfrac{-1}{\sqrt{13}}$.

The four examples of reciprocals mentioned in the previous paragraph result in the following equalities.

$$\sqrt{2} \cdot \frac{1}{\sqrt{2}} = 1$$

$$\sqrt{7} \cdot \frac{1}{\sqrt{7}} = 1$$

$$2\sqrt{21} \cdot \frac{1}{2\sqrt{21}} = 1$$

$$-\sqrt{13} \cdot \frac{-1}{\sqrt{13}} = 1$$

What is to be the result of multiplying $\sqrt{2}$ by the reciprocal of $\sqrt{3}$ or

$\sqrt{2} \cdot \dfrac{1}{\sqrt{3}}$? The answer is to be $\dfrac{\sqrt{2}}{\sqrt{3}}$ where the numerators and denomi-

nators are multiplied in the same way as practiced earlier with rational numbers.

$$\sqrt{2} \cdot \frac{1}{\sqrt{3}} = \frac{\sqrt{2}}{1} \cdot \frac{1}{\sqrt{3}} = \frac{\sqrt{2}}{\sqrt{3}}$$

This procedure, if reversed, gives us an interpretation for radical fractions such as $\dfrac{\sqrt{5}}{\sqrt{11}}$ and $\dfrac{^-\sqrt{13}}{\sqrt{6}}$ .

$$\frac{\sqrt{5}}{\sqrt{11}} \quad \text{means} \quad \sqrt{5} \cdot \frac{1}{\sqrt{11}}$$

$$\frac{^-\sqrt{13}}{\sqrt{6}} \quad \text{means} \quad ^-\sqrt{13} \cdot \frac{1}{\sqrt{6}}$$

Finally we are able to describe the multiplication of two radical fractions such as $\dfrac{\sqrt{3}}{\sqrt{7}} \cdot \dfrac{\sqrt{5}}{\sqrt{11}}$ as necessitating the multiplication of the numerators and the separate multiplication of the denominators.

$$\frac{\sqrt{3}}{\sqrt{7}} \cdot \frac{\sqrt{5}}{\sqrt{11}} = \frac{\sqrt{3} \cdot \sqrt{5}}{\sqrt{7} \cdot \sqrt{11}} = \frac{\sqrt{15}}{\sqrt{77}}$$

Whenever a situation arises in either the numerator or denominator where a radical expression can be simplified, then such a simplification is desirable. Notice the simplification of the numerator in the final answer of the problem below.

$$\frac{\sqrt{6}}{\sqrt{7}} \cdot \frac{\sqrt{2}}{\sqrt{5}} = \frac{\sqrt{6} \cdot \sqrt{2}}{\sqrt{7} \cdot \sqrt{5}}$$

$$= \frac{\sqrt{12}}{\sqrt{35}}$$

$$= \frac{2\sqrt{3}}{\sqrt{35}}$$

EXERCISE 8.9.1

Multiply and simplify whenever possible.

1. $\dfrac{\sqrt{5}}{\sqrt{11}} \cdot \dfrac{\sqrt{2}}{\sqrt{7}}$

2. $\dfrac{\sqrt{6}}{\sqrt{13}} \cdot \dfrac{\sqrt{5}}{\sqrt{7}}$

3. $\dfrac{\sqrt{6}}{\sqrt{15}} \cdot \dfrac{\sqrt{5}}{\sqrt{2}}$

4. $\dfrac{3\sqrt{2}}{\sqrt{5}} \cdot \dfrac{\sqrt{7}}{\sqrt{3}}$

5. $\dfrac{\sqrt{6}}{\sqrt{7}} \cdot \dfrac{\sqrt{2}}{\sqrt{11}}$

8. $\dfrac{\sqrt{5}}{\sqrt{2}} \cdot \dfrac{\sqrt{7}}{\sqrt{2}}$

6. $\dfrac{\sqrt{7}}{\sqrt{2}} \cdot \dfrac{\sqrt{10}}{\sqrt{35}}$

9. $\dfrac{\sqrt{6}}{\sqrt{7}} \cdot \dfrac{\sqrt{6}}{\sqrt{7}}$

7. $\dfrac{\sqrt{6}}{\sqrt{7}} \cdot \dfrac{\sqrt{5}}{\sqrt{7}}$

10. $\dfrac{\sqrt{3}}{\sqrt{2}} \cdot \dfrac{\sqrt{2}}{\sqrt{3}}$

## 8.10  Simplifying Radical Fractions

Every radical fraction with counting numbers as radicands can be written as a single radical with a fractional radicand. For example,

$\dfrac{\sqrt{10}}{\sqrt{15}}$ is a radical fraction in which each of the radicands is a counting

number. This fraction can be written as $\sqrt{\dfrac{10}{15}}$ where there is only one

radical sign and the radicand is a rational number.

$$\dfrac{\sqrt{10}}{\sqrt{15}} \text{ is equivalent to } \sqrt{\dfrac{10}{15}} \, .$$

Every radical expression with a positive rational radicand can be written as a radical fraction in which both the numerator and denominator

have counting numbers as their radicands. For example, $\sqrt{\dfrac{6}{9}}$ can be

written as $\dfrac{\sqrt{6}}{\sqrt{9}}$ .

The previous two paragraphs provide the opportunity to write any radical fraction using either one or two radical signs. This flexibility is most helpful in the simplification of radical fractions.

The final result of simplifying a radical fraction should be a fraction in which the denominator is a counting number not a radical expression. Furthermore, the numerator and denominator should have no common integer factors other than 1 and $^-1$.

Two basic procedures are necessary in simplifying radical fractions. The first is the successful elimination of common factors from numerator and denominator. This simplification is shown next where a single radical sign is all that is necessary.

$$\frac{\sqrt{10}}{\sqrt{15}} = \sqrt{\frac{10}{15}} = \sqrt{\frac{2}{3}}$$

The second procedure provides a denominator that is a perfect square and, making use of the fact that two radical signs may be used, an integer is obtained for the denominator of the simplified fraction.

$$\sqrt{\frac{2}{3}} = \sqrt{\frac{2}{3} \cdot \frac{3}{3}} = \sqrt{\frac{6}{9}} = \frac{\sqrt{6}}{\sqrt{9}} = \frac{\sqrt{6}}{3}$$

Note that any rational number may be written with a perfect square integer as its denominator by multiplying the rational number by an appropriately chosen symbol for 1. In the case of $\frac{2}{3}$ we multiply by $\frac{3}{3}$ where the choice of 3's was identical to the denominator of $\frac{2}{3}$. This possibility will always exist, but is sometimes unnecessary. For example, to write $\frac{3}{8}$ with a perfect square denominator it is sufficient to choose $\frac{2}{2}$ to multiply by, since 16 is a perfect square integer.

$$\frac{\sqrt{15}}{\sqrt{40}} = \sqrt{\frac{15}{40}}$$

$$= \sqrt{\frac{3}{8}}$$

$$= \sqrt{\frac{3}{8} \cdot \frac{2}{2}}$$

$$= \sqrt{\frac{6}{16}}$$

$$= \frac{\sqrt{6}}{\sqrt{16}}$$

$$= \frac{\sqrt{6}}{4}$$

Below is shown the simplification of a more involved radical fraction.

$$\frac{-5\sqrt{6}}{3\sqrt{15}} = \frac{-5}{3} \cdot \sqrt{\frac{6}{15}}$$

$$= \frac{-5}{3} \sqrt{\frac{2}{5}}$$

$$= \frac{-5}{3} \sqrt{\frac{2}{5} \cdot \frac{5}{5}}$$

$$= \frac{-5}{3} \cdot \frac{\sqrt{10}}{\sqrt{25}}$$

$$= \frac{-5}{3} \cdot \frac{\sqrt{10}}{5}$$

$$= \frac{-\sqrt{10}}{3}$$

EXERCISE 8.10.1

Simplify:

1. $\sqrt{\dfrac{3}{5}}$

2. $\sqrt{\dfrac{2}{3}}$

3. $\sqrt{\dfrac{4}{7}}$

4. $\sqrt{\dfrac{2}{9}}$

5. $\sqrt{\dfrac{6}{11}}$

6. $\sqrt{\dfrac{21}{2}}$

7. $\sqrt{\dfrac{7}{10}}$

8. $\sqrt{\dfrac{5}{3}}$

9. $\sqrt{\dfrac{10}{7}}$

10. $\sqrt{\dfrac{3}{13}}$

11. $\dfrac{\sqrt{6}}{\sqrt{8}}$

12. $\dfrac{\sqrt{15}}{\sqrt{30}}$

13. $\dfrac{\sqrt{10}}{\sqrt{6}}$

22. $\dfrac{2\sqrt{7}}{^-\sqrt{8}}$

14. $\dfrac{\sqrt{14}}{\sqrt{22}}$

23. $\dfrac{6\sqrt{3}}{5\sqrt{10}}$

15. $\dfrac{\sqrt{5}}{\sqrt{15}}$

24. $\dfrac{^-2\sqrt{3}}{5\sqrt{6}}$

16. $\dfrac{\sqrt{14}}{\sqrt{35}}$

25. $\dfrac{7\sqrt{15}}{8\sqrt{35}}$

17. $\dfrac{\sqrt{30}}{\sqrt{55}}$

26. $\dfrac{^-6\sqrt{6}}{5\sqrt{10}}$

18. $\dfrac{\sqrt{35}}{\sqrt{15}}$

27. $\dfrac{^-9\sqrt{5}}{^-2\sqrt{12}}$

19. $\dfrac{\sqrt{39}}{\sqrt{26}}$

28. $\dfrac{6\sqrt{14}}{7\sqrt{6}}$

20. $\dfrac{\sqrt{55}}{\sqrt{77}}$

29. $\dfrac{2\sqrt{33}}{5\sqrt{55}}$

21. $\dfrac{4\sqrt{2}}{5\sqrt{3}}$

30. $\dfrac{^-4\sqrt{30}}{5\sqrt{21}}$

## 8.11  Addition of Radical Expressions

Before explaining the addition of radical expressions, two situations that were presented earlier are worthy of a brief review.

Recall that the numerical expression $4 + 5 \cdot 3$ has an evaluation of 19 because the multiplication of $5 \cdot 3$ must be completed before any addition. In the expression $4 + 5\sqrt{3}$ a similar situation exists. $5\sqrt{3}$ means $5 \cdot \sqrt{3}$. Consequently, the multiplication precedes the addition in $4 + 5\sqrt{3}$ and it would be incorrect to add $4 + 5$ in simplifying $4 + 5\sqrt{3}$.

There is one situation wherein the order of operations, multiplication before addition, may be altered. This situation is the substance of the Distributive Law of Multiplication over Addition. Algebraic binomials such as $5x + 3x$ and $2x^2y - 7x^2y$ can be simplified to monomials because the terms have common variable factors and upon factoring the resultant binomial is a numerical expression.

$$5x + 3x = (5 + 3)x = 8x$$

$$2x^2y - 7x^2y = (2 - 7)x^2y = {}^-5x^2y$$

Compare the two previous examples with the situation posed by $4x + 5y$ or $9x^2y - 7xy^2$. These binomials can not be simplified to monomials. The reason for $4x + 5y$ is simply that the two terms have no common variable factors. The reason for $9x^2y - 7xy^2$ is that after the common variable factors have been found the resultant binomial is still not a numerical expression and defies further simplification.

$$9x^2y - 7xy^2 = (9x - 7y)xy$$

Now consider the following addition expression involving radicals:

$$7 + 3\sqrt{5} - 2\sqrt{7} + 4 - 6\sqrt{7} - \sqrt{5}$$

There are six terms to the expression. Looking at the first two terms, $7 + 3\sqrt{5}$, observe that $3\sqrt{5}$ means $3 \cdot \sqrt{5}$ and therefore the multiplication must precede the addition. $7 + 3\sqrt{5}$ can not be simplified. Similarly, the second and third terms, $3\sqrt{5} - 2\sqrt{7}$, can not be simplified because the multiplication must be completed first unless there is a common factor and these terms have no common factors other than 1 and $^-1$.

But the six-term expression can be simplified. The commutative and associative properties of addition allow the order and grouping of the terms in an addition expression to be altered. The first step in simplifying the six-term expression is shown below.

$$7 + 3\sqrt{5} - 2\sqrt{7} + 4 - 6\sqrt{7} - \sqrt{5}$$

$$(7 + 4) + (3\sqrt{5} - \sqrt{5}) + ({}^-2\sqrt{7} - 6\sqrt{7})$$

The expression $(7 + 4)$ offers no difficulty in being simplified to 11. The expression $(3\sqrt{5} - \sqrt{5})$ is simplified using the fact that $\sqrt{5}$ is a factor of both terms.

$$3\sqrt{5} - \sqrt{5} = 3\sqrt{5} - 1\sqrt{5} = (3 - 1) \cdot \sqrt{5} = 2\sqrt{5}$$

$({}^-2\sqrt{7} - 6\sqrt{7})$ is also simplified using the distributive property and the fact that $\sqrt{7}$ is a common factor.

$${}^-2\sqrt{7} - 6\sqrt{7} = ({}^-2 - 6) \cdot \sqrt{7} = {}^-8\sqrt{7}$$

Hence the complete simplification of the addition expression is shown below.

$$7 + 3\sqrt{5} - 2\sqrt{7} + 4 - 6\sqrt{7} - \sqrt{5}$$

$$(7 + 4) + (3\sqrt{5} - \sqrt{5}) + (^-2\sqrt{7} - 6\sqrt{7})$$

$$11 + 2\sqrt{5} - 8\sqrt{7}$$

Note the reasons the preceding example could not be further simplified.

As another example of the simplification of addition expressions study the following problem and the manner in which it is handled.

$$6\sqrt{7} - 2\sqrt{13} + 4 + 5\sqrt{13} - 2\sqrt{6} + 3\sqrt{7}$$

$$(6\sqrt{7} + 3\sqrt{7}) + (^-2\sqrt{13} + 5\sqrt{13}) + 4 - 2\sqrt{6}$$

$$(6 + 3) \cdot \sqrt{7} + (^-2 + 5) \cdot \sqrt{13} + 4 - 2\sqrt{6}$$

$$9\sqrt{7} + 3\sqrt{13} + 4 - 2\sqrt{6}$$

The example above shows both when and how simplification of addition expressions is accomplished.

EXERCISE 8.11.1

Simplify:

1. $6 - 3\sqrt{2} + 5 + 8\sqrt{2}$
2. $2\sqrt{7} + 7\sqrt{2} - 6\sqrt{2} + \sqrt{7}$
3. $5\sqrt{10} - 2\sqrt{13} - \sqrt{10} + 2\sqrt{13}$
4. $3\sqrt{5} - 2 + 5\sqrt{7} - 6$
5. $^-2\sqrt{11} + 5 - 4\sqrt{11} + 6 - \sqrt{11}$
6. $6 - 3\sqrt{2} - 4\sqrt{5} - 3 + \sqrt{5} - 5\sqrt{2}$
7. $4\sqrt{5} - 8\sqrt{19} + 4\sqrt{6} - 3\sqrt{5} + \sqrt{6}$
8. $5 + 7\sqrt{6} + 4\sqrt{3} - 3\sqrt{2} + 2\sqrt{3}$
9. $6 - 8 + 4\sqrt{3} - 6\sqrt{7} + \sqrt{7} + 5\sqrt{3}$
10. $9 + 3\sqrt{15} + 2\sqrt{11} - 3 - 4\sqrt{11} - 2\sqrt{15}$
11. $5 - 7\sqrt{10} + 2\sqrt{6} - 4\sqrt{10} - 1$
12. $2\sqrt{31} - 6\sqrt{19} + 2\sqrt{19} + 7\sqrt{6} - 2\sqrt{31}$
13. $5\sqrt{3} + 4\sqrt{3} - 2\sqrt{5} - 2\sqrt{3} - 4\sqrt{5}$
14. $4\sqrt{22} - 2\sqrt{85} + 6\sqrt{2} - 3\sqrt{85} + 4\sqrt{2}$
15. $9\sqrt{3} + 2\sqrt{6} + 5\sqrt{17} - 3\sqrt{6}$
16. $4\sqrt{19} - 13\sqrt{13} + 6\sqrt{19} + 2\sqrt{19} + \sqrt{2}$
17. $5\sqrt{6} - \sqrt{29} + \sqrt{6} - \sqrt{29} + 3\sqrt{7}$

18.  $-2\sqrt{11} + 8\sqrt{3} - 2\sqrt{5} + \sqrt{11} - \sqrt{5}$
19.  $3\sqrt{2} + 5 - 6\sqrt{2} + 7 - 3 + 2\sqrt{2}$
20.  $4\sqrt{71} - 2\sqrt{39} + 6\sqrt{39} + 5\sqrt{39} - 1$

A student who likes to formulate rules for attacking problems might easily have come to the conclusion that you can simplify the addition of two radical expressions only when the radicands are the same. This is essentially true, but each radical expression should be simplified before applying such a criterion.

Consider the following simplification:

$$8\sqrt{2} - 5\sqrt{18}$$

Since $\sqrt{18}$ can be simplified the following steps are used:

$$8\sqrt{2} - 5\sqrt{18}$$
$$8\sqrt{2} - 5\sqrt{9}\cdot\sqrt{2}$$
$$8\sqrt{2} - 5\cdot3\cdot\sqrt{2}$$
$$8\sqrt{2} - 15\sqrt{2}$$
$$(8 - 15)\cdot\sqrt{2}$$
$$-7\sqrt{2}$$

EXERCISE 8.11.2

Simplify each addition expression by first simplifying each radical whenever possible.

1.  $5\sqrt{3} + 3\sqrt{12}$
2.  $2\sqrt{20} - 7\sqrt{5}$
3.  $7\sqrt{8} - \sqrt{2}$
4.  $\sqrt{36} - 5\sqrt{3} - \sqrt{27} + 3\sqrt{4}$
5.  $6\sqrt{2} - \sqrt{18}$
6.  $5\sqrt{6} - \sqrt{54}$
7.  $4\sqrt{7} + 5\sqrt{28}$
8.  $6\sqrt{9} - 4\sqrt{5} - \sqrt{20} - \sqrt{81}$
9.  $5\sqrt{44} + 2\sqrt{11}$
10. $9\sqrt{72} + \sqrt{2}$

11. $\sqrt{8} + \sqrt{32}$
12. $\sqrt{40} + \sqrt{90}$
13. $\sqrt{12} - \sqrt{147}$
14. $\sqrt{18} + \sqrt{162}$
15. $3\sqrt{52} - 6\sqrt{13}$
16. $5\sqrt{80} - \sqrt{45}$
17. $7\sqrt{27} - 6\sqrt{12}$
18. $4\sqrt{48} - \sqrt{108}$
19. $5\sqrt{128} + \sqrt{200}$
20. $2\sqrt{300} - 7\sqrt{75}$

## 8.12 Solving Linear Equations with Irrational Solutions

$x\sqrt{2} - \sqrt{12} = \sqrt{3}$ is a linear equation. A method of solving this equation is very similar to the procedure used on equations such as $5x - 17 = 6$.

Recall that $5x - 17 = 6$ could be solved by adding the opposite of $^-17$ to both sides of the equation and then, after simplifying, multiplying both sides of the equation by the reciprocal of 5.

$$5x - 17 = 6$$

$$(5x - 17) + 17 = 6 + 17$$

$$5x = 23$$

$$\frac{1}{5} \cdot 5x = \frac{1}{5} \cdot 23$$

$\left\{\dfrac{23}{5}\right\}$ is the truth set of $5x - 17 = 6$.

The same procedure of adding opposites and multiplying by reciprocals is used to solve $x\sqrt{2} - \sqrt{12} = \sqrt{3}$.

$$x\sqrt{2} - \sqrt{12} = \sqrt{3}$$

$$(x\sqrt{2} - \sqrt{12}) + \sqrt{12} = \sqrt{3} + \sqrt{12}$$

$$x\sqrt{2} = \sqrt{3} + 2\sqrt{3}$$

$$x\sqrt{2} = 3\sqrt{3}$$

$$\frac{1}{\sqrt{2}} \cdot x\sqrt{2} = \frac{1}{\sqrt{2}} \cdot 3\sqrt{3}$$

$$x = \frac{3\sqrt{3}}{\sqrt{2}}$$

$$x = \frac{3\sqrt{6}}{2}$$

$\left\{\dfrac{3\sqrt{6}}{2}\right\}$ is the truth set of $x\sqrt{2} - \sqrt{12} = \sqrt{3}$.

As another example, the solution of $4x - \sqrt{7} = \sqrt{5}$ is worked out below.

$$4x - \sqrt{7} = \sqrt{5}$$

$$(4x - \sqrt{7}) + \sqrt{7} = \sqrt{5} + \sqrt{7}$$

$$4x = \sqrt{5} + \sqrt{7}$$

$$\frac{1}{4} \cdot 4x = \frac{1}{4}(\sqrt{5} + \sqrt{7})$$

$$x = \frac{\sqrt{5} + \sqrt{7}}{4}$$

$\left\{ \dfrac{\sqrt{5} + \sqrt{7}}{4} \right\}$ is the truth set of $4x - \sqrt{7} = \sqrt{5}$.

EXERCISE 8.12.1

Find truth sets using the set of real numbers as the domain.

1. $x - 3 = \sqrt{5}$
2. $x + 5 = {}^{-}\sqrt{8}$
3. $x - \sqrt{7} = \sqrt{25}$
4. $x - \sqrt{12} = \sqrt{27}$
5. $x\sqrt{5} = {}^{-}3\sqrt{2}$
6. $x\sqrt{8} = \sqrt{10}$
7. $x\sqrt{3} + \sqrt{20} = 4\sqrt{5}$
8. $x\sqrt{7} - \sqrt{6} = \sqrt{54}$
9. $5x - \sqrt{2} = \sqrt{6}$
10. $2x - \sqrt{10} = 5$

11. $x\sqrt{16} - \sqrt{6} = \sqrt{11}$
12. $3x - 2 = \sqrt{13}$
13. $9x + \sqrt{27} = \sqrt{48}$
14. $x\sqrt{2} + \sqrt{7} = \sqrt{28}$
15. $x\sqrt{6} - \sqrt{48} = \sqrt{27}$
16. $x\sqrt{5} + 4\sqrt{3} = 2\sqrt{12}$
17. $x\sqrt{9} - 2\sqrt{20} = 7\sqrt{162}$
18. $x\sqrt{4} + 6\sqrt{5} = 2\sqrt{200}$
19. $x\sqrt{8} - 2 = 7 + x\sqrt{2}$
20. $x\sqrt{3} - \sqrt{6} = \sqrt{24} - x\sqrt{12}$

### 8.13   Solving Equations of the Form $(x + a)^2 = b$

$x^2 = 16$ is a simple quadratic equation. To solve it, the main question to answer is "What numbers can be squared to give a product of 16?" There are two numbers, one positive and one negative, which when squared give 16. Hence, the truth set of $x^2 = 16$ is $\{4, {}^{-}4\}$.

The truth set of $x^2 = 7$ is possibly not as obvious as the truth set of $x^2 = 16$, but it requires the same type of question. What numbers can be squared to give 7? Again there are two such numbers, one positive and one negative, but these numbers are not integers. The truth set is $\{\sqrt{7}, {}^{-}\sqrt{7}\}$ since the square of either of these numbers is 7.

The equation $(x + 1)^2 = 25$ presents a slight extension of the thought involved in solving $x^2 = 16$. In $(x + 1)^2 = 25$ it is the quantity $(x + 1)$ that is squared and, therefore, the numbers that can be squared to give 25 must replace $(x + 1)$ rather than just $x$. Since $(^-5)^2 = 25$ and $(5)^2 = 25$ then $x + 1 = ^-5$ or $x + 1 = 5$. Solving these two equations produces the truth set of $\{4, ^-6\}$. The student should check both of these numbers in $(x + 1)^2 = 25$ to see that they produce true statements.

The equation $(x - 4)^2 = 11$ asks that the numbers that can be squared to give 11 be used to replace the quantity $(x - 4)$. Since $(\sqrt{11})^2 = 11$ and $(^-\sqrt{11})^2 = 11$ then these values must replace $(x - 4)$. This line of thought provides two linear equations to be solved:

$$x - 4 = \sqrt{11} \qquad x - 4 = ^-\sqrt{11}$$

The solutions of these equations are $4 + \sqrt{11}$ and $4 - \sqrt{11}$ and, therefore, the truth set of $(x - 4)^2 = 11$ is $\{4 + \sqrt{11}, 4 - \sqrt{11}\}$.

EXERCISE 8.13.1

Find truth sets using the set of real numbers as the domain.

1.  $x^2 = 49$
2.  $x^2 = 100$

3.  $x^2 = ^-16$   (Be careful!)
4.  $x^2 = 41$
5.  $x^2 = 19$
6.  $(x - 1)^2 = 4$
7.  $(x + 3)^2 = 64$
8.  $(x + 5)^2 = 36$
9.  $(x - 2)^2 = ^-9$
10. $(x - 1)^2 = 13$

11. $(x + 7)^2 = 6$
12. $(x - 3)^2 = 8$
    (Simplify the answer.)
13. $(x + 6)^2 = ^-5$
14. $(x - 2)^2 = 12$
15. $(2x - 1)^2 = 7$
16. $(4x - 9)^2 = 10$
17. $(3x + 1)^2 = 24$
18. $(5x + 3)^2 = 60$
19. $(3x - 10)^2 = ^-31$
20. $(6x - 5)^2 = 42$

## 8.14  Perfect Square Trinomials

The equation $(x - 3)^2 = 7$ is solved by using the fact that $(\sqrt{7})^2 = 7$ and $(^-\sqrt{7})^2 = 7$ and, therefore, $(x - 3)$ must be equal to either $\sqrt{7}$ or $^-\sqrt{7}$ to provide a true statement from $(x - 3)^2 = 7$. From these facts two linear equations are obtained and the union of their truth sets is the truth set of $(x - 3)^2 = 7$.

$$x - 3 = \sqrt{7} \qquad x - 3 = ^-\sqrt{7}$$
$$x = 3 + \sqrt{7} \qquad x = 3 - \sqrt{7}$$

$\{3 + \sqrt{7}, 3 - \sqrt{7}\}$ is the truth set of $(x - 3)^2 = 7$.

This equation solving procedure was only possible because the left side of the equation, $(x - 3)^2 = 7$, is a quantity squared and that quantity must be either the positive or negative square root of the right side of the equality to provide a true statement.

Suppose the equation $x^2 - 10x + 25 = 11$ were given. Factoring $x^2 - 10x + 25$ we obtain $(x - 5)(x - 5)$ or $(x - 5)^2$. Hence the given equation is equivalent to $(x - 5)^2 = 11$ and may be solved in the same manner as those in the last exercise.

What the preceding paragraph shows is that we now could solve any quadratic equation if we could be assured of factoring a polynomial that is a perfect square for one member of the equation. This is always possible and a discussion for producing such perfect square trinomials is the intent of this section.

Below are shown four perfect square trinomials. What is the number relationship between the second and third terms in each trinomial?

$$x^2 - 6x + 9$$
$$x^2 + 2x + 1$$
$$x^2 - 12x + 36$$
$$x^2 + 8x + 16$$

From the chapter on factoring we know that the coefficient of the middle term must be the sum of the factors of the third term. When the trinomial is a perfect square then the coefficient of the middle term must be twice the square root of the third term.

Using $x^2 + bx + c$ as a general form for a trinomial perfect square we can represent the relationship between the middle and third terms by the equality (formula) $\left(\frac{1}{2} \cdot b\right)^2 = c$. Half the middle term squared will give the third term.

Let us try the formula in attempting to make a perfect square trinomial from $x^2 - 14x +$ ____. The middle term's coefficient, $b$, is $^-14$. Using the formula $\left(\frac{1}{2} \cdot b\right)^2 = c$,

$$\left(\frac{1}{2} \cdot {}^-14\right)^2 = c$$
$$({}^-7)^2 = c$$
$$49 = c$$

Using 49 as the third term for $x^2 - 14x + \underline{\quad}$ we arrive at $x^2 - 14x + 49$ which is a perfect square trinomial because it is factored as $(x - 7)^2$. Note that the $^-7$ in the factorization is $\left(\dfrac{1}{2} \cdot b\right)$.

As another example, consider the binomial $x^2 + 5x$. According to the formula for finding the third term of a perfect square trinomial, we replace $b$ by 5 in $\left(\dfrac{1}{2} \cdot b\right)^2 = c$ and

$$\left(\frac{1}{2} \cdot 5\right)^2 = c$$

$$\left(\frac{5}{2}\right)^2 = c$$

$$\frac{25}{4} = c$$

The trinomial we obtained is $x^2 + 5x + \dfrac{25}{4}$ and it is a perfect square trinomial because it is factored as $\left(x + \dfrac{5}{2}\right)^2$. This factorization may be checked by multiplication and it is suggested that you do so. Notice that the $\dfrac{5}{2}$ in the factorization is $\left(\dfrac{1}{2} \cdot b\right)$.

One last example of constructing a perfect square trinomial will be given. Suppose we start with $x^2 - \dfrac{4}{3}x$. By the formula $\left(\dfrac{1}{2} \cdot b\right)^2 = c$ and using $b = \dfrac{^-4}{3}$,

$$\left(\frac{1}{2} \cdot \frac{^-4}{3}\right)^2 = c$$

$$\left(\frac{^-2}{3}\right)^2 = c$$

$$\frac{4}{9} = c$$

Hence the desired trinomial is $x^2 - \dfrac{4}{3}x + \dfrac{4}{9}$ which is factored as $\left(x - \dfrac{2}{3}\right)^2$.

Once again, the $\dfrac{-2}{3}$ in the factorization is $\left(\dfrac{1}{2} \cdot b\right)$.

EXERCISE 8.14.1

In each problem a binomial is given. Find a third term to complete a perfect square trinomial and then factor it.

| | | | |
|---|---|---|---|
| 1. | $x^2 + 6x$ | 11. | $x^2 + 11x$ |
| 2. | $x^2 - 18x$ | 12. | $x^2 + 24x$ |
| 3. | $x^2 + 22x$ | 13. | $x^2 - x$ |
| 4. | $x^2 - 8x$ | 14. | $x^2 + 13x$ |
| 5. | $x^2 + 7x$ | 15. | $x^2 - 40x$ |
| 6. | $x^2 - 9x$ | 16. | $x^2 + \dfrac{5}{2}x$ |
| 7. | $x^2 + x$ | 17. | $x^2 - \dfrac{7}{3}x$ |
| 8. | $x^2 - 3x$ | 18. | $x^2 + \dfrac{1}{3}x$ |
| 9. | $x^2 + 4x$ | 19. | $x^2 - \dfrac{2}{3}x$ |
| 10. | $x^2 - 16x$ | 20. | $x^2 - \dfrac{4}{5}x$ |

## 8.15  Solving Quadratic Equations by Completing the Square

Recall that the equation $(x - 3)^2 = 6$ is solved by writing the two linear equations $x - 3 = \sqrt{6}$ and $x - 3 = {}^-\sqrt{6}$. Solving these two linear equations gives the truth set of $\{3 + \sqrt{6}, 3 - \sqrt{6}\}$.

Now consider the equation $x^2 - 7x = 5$. For this equation the left member, $x^2 - 7x$, is not now a perfect square, but we can make it one using the method of the last exercise.

$$x^2 - 7x \qquad \left(\frac{1}{2} \cdot {}^-7\right)^2 = c$$

$$\left(\frac{{}^-7}{2}\right)^2 = c$$

$$\frac{49}{4} = c$$

$x^2 - 7x + \dfrac{49}{4}$ is the perfect square trinomial.

Now back to solving the quadratic equation $x^2 - 7x = 5$. Any number may be added to each side of any equation. Since a perfect square trinomial is desired as the left member, $\dfrac{49}{4}$ is added to both members of $x^2 - 7x = 5$ in the second step of the example shown below. Study the solution of $x^2 - 7x = 5$; this same method may be applied to any quadratic equation.

$$x^2 - 7x = 5$$

$$x^2 - 7x + \frac{49}{4} = 5 + \frac{49}{4}$$

$$\left(x - \frac{7}{2}\right)^2 = \frac{69}{4}$$

$$x - \frac{7}{2} = \sqrt{\frac{69}{4}} \qquad\qquad x - \frac{7}{2} = {}^-\sqrt{\frac{69}{4}}$$

$$x - \frac{7}{2} = \frac{\sqrt{69}}{2} \qquad\qquad x - \frac{7}{2} = \frac{{}^-\sqrt{69}}{2}$$

$$x = \frac{7}{2} + \frac{\sqrt{69}}{2} \qquad\qquad x = \frac{7}{2} - \frac{\sqrt{69}}{2}$$

The truth set of $x^2 - 7x = 5$ is $\left\{ \dfrac{7 + \sqrt{69}}{2}, \dfrac{7 - \sqrt{69}}{2} \right\}$.

Quadratic equations such as $x^2 - 9x + 8 = 0$ and $x^2 + 7x + 12 = 0$ should still be solved by the factoring method taught in chapter 7. Whenever the elements of the truth set of a quadratic equation are rational numbers then the factoring method is a simpler way for finding truth sets.

When given an equation like $x^2 = 9x - 4$ the first step in its solution should be an attempt to solve by the factoring method. If that attempt fails then proceed to the construction of a perfect square trinomial.

The equation $x^2 = 9x - 4$ may be solved in the following steps:

$$x^2 = 9x - 4$$

$$x^2 - 9x + 4 = 0$$

$$(x^2 - 9x + 4) \quad \text{is prime}$$

$$x^2 - 9x = {}^-4$$

$$x^2 - 9x + \frac{81}{4} = {}^-4 + \frac{81}{4} \qquad \left(\frac{1}{2} \cdot {}^-9\right)^2 = \frac{81}{4}$$

$$\left(x - \frac{9}{2}\right)^2 = \frac{65}{4}$$

$$x - \frac{9}{2} = \sqrt{\frac{65}{4}} \qquad\qquad x - \frac{9}{2} = {}^-\sqrt{\frac{65}{4}}$$

$$x = \frac{9}{2} + \frac{\sqrt{65}}{2} \qquad\qquad x = \frac{9}{2} - \frac{\sqrt{65}}{2}$$

$$\left\{\frac{9 + \sqrt{65}}{2}, \frac{9 - \sqrt{65}}{2}\right\} \text{ is the truth set of } x^2 = 9x - 4$$

Whenever a quadratic equation is being solved and the perfect square trinomial is equal to a negative number, then the equation has the empty set as its truth set. $(x - 5)^2 = {}^-2$ and $\left(x + \frac{3}{2}\right)^2 = {}^-7$ are examples of such situations. Since the square of a real number can not be negative the truth set is the empty set.

EXERCISE 8.15.1

Find the truth set using the set of real numbers as the domain.

|     |                      |     |                      |
|-----|----------------------|-----|----------------------|
| 1.  | $x^2 + 12x = {}^-17$ | 6.  | $x^2 + 7x = 3$       |
| 2.  | $x^2 - 8x = 19$      | 7.  | $x^2 - x = 9$        |
| 3.  | $x^2 + 5x = 6$       | 8.  | $x^2 - 10x = {}^-8$  |
| 4.  | $x^2 - 4x = {}^-3$   | 9.  | $x^2 - 2x = 7$       |
| 5.  | $x^2 - 6x = 12$      | 10. | $x^2 - 5x = 1$       |

11.  $x^2 + 4x = {}^-13$          16.  $x^2 = 3 - 6x$
12.  $x^2 = 2x + 5$          17.  $x^2 = 2x - 5$
13.  $x^2 = 6x - 3$          18.  $x^2 - 9 = 2x$
14.  $x^2 - 9 = 5x$          19.  $x^2 = 4 - 3x$
15.  $x^2 + 4 = 5x$          20.  $x^2 + x = 6$

## 8.16   The Quadratic Formula

The method used in the last exercise to solve quadratic equations is called *completing the square*. It has advantages and disadvantages. Among its advantages is the fact that it may be used to solve any quadratic equation whether the truth set contains rational or irrational numbers. It also clearly shows when no real number is a solution of a quadratic equation.

The ability of the *completing the square* method to solve any quadratic equation makes it a powerful, though at times cumbersome, tool in equation solving.

In this section the power of the *completing the square* method is used to develop an easier method for solving all quadratic equations. A formula will be found that shows all solutions of a quadratic equation in terms of the numbers that appear in the equation. This formula is called the *quadratic formula*.

Before developing the formula one more example of solving using the completing the square method will be shown. The equation to be solved is $3x^2 - 7x - 2 = 0$. This equation is slightly more involved than those of the last exercise because the first term, $3x^2$, has a coefficient of 3 whereas those solved earlier all had coefficients of 1 for the first term.

For the solution of $3x^2 - 7x - 2 = 0$ the first step is to multiply both members of the equation by $\frac{1}{3}$. From that point on the equation is solved like those of the last exercise. Study this example carefully. It is important to understand the necessity for each step.

$$3x^2 - 7x - 2 = 0$$

$$\frac{1}{3}(3x^2 - 7x - 2) = \frac{1}{3} \cdot 0$$

$$x^2 - \frac{7}{3}x - \frac{2}{3} = 0$$

$$x^2 - \frac{7}{3}x = \frac{2}{3}$$

$$x^2 - \frac{7}{3}x + \frac{49}{36} = \frac{2}{3} + \frac{49}{36} \qquad \left(\frac{1}{2} \cdot \frac{^-7}{3}\right)^2 = \frac{49}{36}$$

$$\left(x - \frac{7}{6}\right)^2 = \frac{73}{36} \qquad \left(\frac{1}{2} \cdot \frac{^-7}{3}\right) = \frac{^-7}{6}$$

$$x - \frac{7}{6} = \sqrt{\frac{73}{36}} \qquad x - \frac{7}{6} = {}^-\sqrt{\frac{73}{36}}$$

$$x = \frac{7 + \sqrt{73}}{6} \qquad x = \frac{7 - \sqrt{73}}{6}$$

$\left\{\dfrac{7 + \sqrt{73}}{6}, \dfrac{7 - \sqrt{73}}{6}\right\}$ is the truth set of $3x^2 - 7x - 2 = 0$.

If each step of the preceding example is clear then it is possible to use exactly the same procedure to solve $ax^2 + bx + c = 0$. This equation has the letters $a$, $b$, and $c$ instead of numbers, but if the equation can be solved in terms of these letters then a formula will have been found that will solve the equation for any real number replacements of $a$, $b$, and $c$.

The solution given below parallels closely the solution of the previous example. If any step is not clear, refer to the solution of $3x^2 - 7x - 2 = 0$.

$$ax^2 + bx + c = 0$$

$$\frac{1}{a}(ax^2 + bx + c) = \frac{1}{a} \cdot 0$$

$$x^2 + \frac{b}{a}x + \frac{c}{a} = 0$$

$$x^2 + \frac{b}{a}x = \frac{^-c}{a}$$

$$x^2 + \frac{b}{a}x + \frac{b^2}{4a^2} = \frac{b^2}{4a^2} - \frac{c}{a} \qquad \left(\frac{1}{2} \cdot \frac{b}{a}\right)^2 = \frac{b^2}{4a^2}$$

$$\left(x + \frac{b}{2a}\right)^2 = \frac{b^2 - 4ac}{4a^2} \qquad \left(\frac{1}{2} \cdot \frac{b}{a}\right) = \frac{b}{2a}$$

$$x + \frac{b}{2a} = \sqrt{\frac{b^2 - 4ac}{4a^2}} \qquad x + \frac{b}{2a} = {}^-\sqrt{\frac{b^2 - 4ac}{4a^2}}$$

$$x + \frac{b}{2a} = \frac{\sqrt{b^2 - 4ac}}{2a} \text{*} \qquad\qquad x + \frac{b}{2a} = \frac{^-\sqrt{b^2 - 4ac}}{2a}$$

$$x = \frac{^-b + \sqrt{b^2 - 4ac}}{2a} \qquad\qquad x = \frac{^-b - \sqrt{b^2 - 4ac}}{2a}$$

$\left\{ \dfrac{^-b + \sqrt{b^2 - 4ac}}{2a}, \dfrac{^-b - \sqrt{b^2 - 4ac}}{2a} \right\}$ is the truth set of $ax^2 + bx + c = 0$.

The truth set worked out is for any quadratic equation of the form $ax^2 + bx + c = 0$. If the equation $3x^2 - 7x - 2 = 0$ is to be solved then $a = 3$, $b = {}^-7$, and $c = {}^-2$. Replacing $a$, $b$, and $c$ by these values in

$\dfrac{^-b + \sqrt{b^2 - 4ac}}{2a}$ we obtain:

$$\frac{^-(^-7) + \sqrt{(^-7)^2 - 4 \cdot 3 \cdot {}^-2}}{2 \cdot 3} = \frac{7 + \sqrt{49 + 24}}{6}$$

$$= \frac{7 + \sqrt{73}}{6}$$

This is one of the solutions found earlier in this section for $3x^2 - 7x - 2 = 0$.

To use the formula, $\left\{ \dfrac{^-b + \sqrt{b^2 - 4ac}}{2a}, \dfrac{^-b - \sqrt{b^2 - 4ac}}{2a} \right\}$ , for a quadratic equation, the equation must be written in the form

$$ax^2 + bx + c = 0$$

to arrive at the correct values for $a$, $b$, and $c$.

The equation $3x = 5x^2 - 2$ is equivalent to $^-5x^2 + 3x + 2 = 0$. The second equation is in the proper form for using the quadratic formula and the correct values are $a = {}^-5$, $b = 3$, and $c = 2$.

EXERCISE 8.16.1

For each equation find an equivalent equation in the form

$$ax^2 + bx + c = 0$$

and find $a$, $b$, and $c$ for the proper use of the quadratic formula.

---

* This step is dependent upon $a$ being positive. However, the same final result is obtained when $a$ is negative.

1.  $3x^2 - 2x + 9 = 0$
2.  $4x^2 + 3x - 7 = 0$
3.  $x^2 - 5x + 13 = 0$
4.  $x^2 - x + 12 = 0$
5.  $x^2 + 3x + 1 = 0$
6.  $3x^2 = 4 - 2x$
7.  $7x = 2x^2 - 5$
8.  $35 = 9x - 3x^2$
9.  $4x^2 - 13 = 9x$
10. $2x^2 + 13 = x^2 + 3x$

11. $7x^2 + 9 = 13x + 6$
12. $7x - 3 = 2x + x^2$
13. $x^2 - 7x = 6$
14. $5x^2 - 7 = 0$
15. $6x^2 - 31 = 4x$
16. $3x - x^2 = 4$
17. $5 = 4x^2 - 7x$
18. $6x^2 + 7 = 10x$
19. $2x^2 - 3x - 9 = 0$
20. $5x^2 = 3x - 7$

## 8.17   Solving Quadratic Equations Using the Formula

The truth set of any quadratic equation can be found by using the quadratic formula. Whenever the solutions are rational numbers it is usually still easier to solve by factoring, but the quadratic formula will find any real number solutions, rational or irrational, and also clearly shows when the truth set is the empty set.

To solve $4x^2 = 2x + 3$ first find an equivalent equation of the form $ax^2 + bx + c = 0$.

$$4x^2 = 2x + 3$$

$$4x^2 - 2x - 3 = 0$$

$$(4x^2 - 2x - 3) \quad \text{is prime}$$

$$a = 4, \quad b = {}^-2, \quad c = {}^-3$$

$$\frac{{}^-b + \sqrt{b^2 - 4ac}}{2a} = \frac{{}^-({}^-2) + \sqrt{({}^-2)^2 - 4 \cdot 4 \cdot {}^-3}}{2 \cdot 4}$$

$$\frac{2 + \sqrt{4 + 48}}{8}$$

$$\frac{2 + \sqrt{52}}{8}$$

$$\frac{2 + 2\sqrt{13}}{8}$$

$$\frac{1 + \sqrt{13}}{4}$$

$\dfrac{1 + \sqrt{13}}{4}$ is one solution of $4x^2 - 2x - 3 = 0$. The other solution is

$\dfrac{1 - \sqrt{13}}{4}$ . Notice that the only difference between the two solutions is the

sign preceding the radical. This will always be the case. Consequently the quadratic equation is often shown as

$$x = \frac{-b \pm \sqrt{b^2 - 4ac}}{2a}$$

The double sign preceding the radical is meant to indicate that one solution is obtained using the plus sign and the other solution comes from the use of the minus sign.

Here is another example using the double-signed formula.

$$x^2 + 5x - 2 = 0$$

$$a = 1, \qquad b = 5, \qquad c = {}^{-}2$$

$$x = \frac{-b \pm \sqrt{b^2 - 4ac}}{2a}$$

$$= \frac{{}^{-}5 \pm \sqrt{5^2 - 4 \cdot 1 \cdot {}^{-}2}}{2 \cdot 1}$$

$$= \frac{{}^{-}5 \pm \sqrt{25 + 8}}{2}$$

$$= \frac{{}^{-}5 \pm \sqrt{33}}{2}$$

$\left\{ \dfrac{{}^{-}5 + \sqrt{33}}{2} , \dfrac{{}^{-}5 - \sqrt{33}}{2} \right\}$ is the truth set of $x^2 + 5x - 2 = 0$.

As noted earlier not every quadratic equation has real number solutions and the quadratic formula can be used to determine when that is the case. There is no real number solution for $3x^2 + 2x + 3 = 0$. This is shown in the next example.

$$3x^2 + 2x + 3 = 0$$

$$a = 3, \quad b = 2, \quad c = 3$$

$$x = \frac{^-b \pm \sqrt{b^2 - 4ac}}{2a}$$

$$= \frac{^-2 \pm \sqrt{4 - 36}}{6}$$

$$= \frac{^-2 \pm \sqrt{^-32}}{6}$$

Since the radicand is negative, $^-32$, and no real number squared can be negative the formula shows that the solutions are not real numbers. Hence, using the set of real numbers as the domain, the truth set of $3x^2 + 2x + 3 = 0$ is $\{ \quad \}$.

EXERCISE 8.17.1

Find truth sets using the set of real numbers as the domain. Some of these equations have rational number solutions and are most easily solved by the factoring method, but most of them will require the use of the quadratic formula.

1. $2x^2 + 5x - 1 = 0$
2. $4x^2 + 9x + 3 = 0$
3. $x^2 - x - 12 = 0$
4. $5x^2 - 3x - 2 = 0$
5. $x^2 + 6x - 2 = 0$
6. $4x^2 + 2 = ^-x$
7. $x^2 = 8x - 3$
8. $^-9x = ^-2x^2 - 9$
9. $6x^2 = 7x + 1$
10. $x^2 + 4x = 3$

11. $x^2 + 17x = ^-72$
12. $2x^2 + 3 = 8x$
13. $5x^2 = x + 1$
14. $x^2 + 4x = 8$
15. $^-6x^2 + 7x - 3 = 0$
16. $^-x^2 + 19x - 48 = 0$
17. $2x^2 + 7x = 5$
18. $4x^2 + x + 1 = 0$
19. $7x = 3x^2 + 2$
20. $4x + 6 = x^2$

# 9 ABSOLUTE VALUE EQUATIONS AND INEQUALITIES

## 9.1 Introduction

The absolute value of a real number is the same as the distance of that real number from zero on the real number line. Since our understanding of distance does not allow negative numbers to be lengths, the concept of absolute value always assigns a nonnegative number to any real number.

When we speak of the absolute value of 5, we mean the distance that 5 is from zero. Hence the absolute value of 5 is 5.

When we speak of the absolute value of $^-3$, we mean the distance that $^-3$ is from zero. Hence the absolute value of $^-3$ is 3.

The absolute value of a positive number is that same positive number. The absolute value of any negative number is the opposite of that negative number.

EXERCISE 9.1.1

1. Find the absolute value of each number.
   (a) 7                    (c) $^-9$
   (b) $^-5$                (d) 10

(e) $\dfrac{13}{5}$                          (g) $\pi$

(f) $\dfrac{^-6}{7}$                          (h) $^-\sqrt{2}$

2.  There is only one real number that does not have a positive absolute value. What is it?

### 9.2  Evaluating Absolute Value Expressions

When the absolute value of a number is desired, it is shown by placing vertical parallel lines around the numeral. "| 6 |" is read as "the absolute value of 6." Consequently | 6 | = 6 is a true statement.

"| ⁻13 |" is read as "the absolute value of ⁻13." | ⁻13 | = 13 is a true statement.

The parallel lines used to denote absolute value serve as a grouping symbol whenever they enclose a numerical expression requiring addition and/or multiplication. In such cases the numerical expression should be evaluated before the absolute value is taken. For example, | ⁻3 + 2·5 | is an absolute value expression in which the absolute value bars enclose the numerical expression (⁻3 + 2·5). In such situations the numerical expression should be evaluated first.

$$| {}^-3 + 2 \cdot 5 | = | 7 | = 7$$

Each of the following absolute value expressions has been evaluated correctly by first evaluating the numerical expression within the absolute value bars.

$$| 9 - 5 | = | 4 | = 4$$
$$| {}^-9 - 5 | = | {}^-14 | = 14$$
$$| 6 - 13 | = | {}^-7 | = 7$$
$$| {}^-6 + 13 | = | 7 | = 7$$

EXERCISE 9.2.1

1.  Evaluate:

(a) | 4 − 6 |
(b) | 13 − 10 |
(c) | 2·⁻5 − 2 |
(d) | 4·⁻2 + 3·5 |
(e) | $\sqrt{2}$ − $\sqrt{8}$ |

(f)  $|\ 2 - 6 \cdot 3\ |$

(g)  $|\ 4 \cdot 5 - 6 \cdot 5\ |$

(h)  $|\ \sqrt{27} - \sqrt{48}\ |$

(i)  $\left|\ \dfrac{3}{5} - \dfrac{4}{7}\ \right|$

(j)  $\left|\ \dfrac{7}{11} - \dfrac{2}{3}\ \right|$

2.  Determine the truth or falsity of each numerical statement.

(a)  $|\ 6 - 4\ | = |\ 6\ | - |\ 4\ |$

(b)  $|\ 6 - 4\ | = |\ 4 - 6\ |$

(c)  $|\ 3 \cdot {}^{-}5\ | = |\ 3\ | \cdot |\ {}^{-}5\ |$

(d)  $|\ 9 - 13\ | = |\ 9\ | - |\ 13\ |$

(e)  $|\ 2 + 7\ | = |\ {}^{-}2 - 7\ |$

(f)  $|\ {}^{-}2 \cdot {}^{-}6\ | = |\ {}^{-}2\ | \cdot |\ {}^{-}6\ |$

(g)  $|\ {}^{-}14 + 2\ | = |\ 14 - 2\ |$

(h)  $|\ {}^{-}3 + {}^{-}2\ | = |\ {}^{-}3\ | + |\ {}^{-}2\ |$

(i)  $|\ {}^{-}6\ | = {}^{-}1 \cdot {}^{-}6$

(j)  $|\ {}^{-}13\ | = {}^{-}1 \cdot {}^{-}13$

## 9.3   Removing Absolute Value Symbols from Open Expressions

The absolute value of any positive real number is that same positive number.

The absolute value of any negative real number is its opposite.

Look at the following open sentence. Will this open sentence become a true numerical statement for all real number replacements of $x$?

$$|\ x\ | = x$$

The answer to the preceding question is a resounding "No." $x$ is a variable and accordingly is neither positive nor negative. The sentence $|\ x\ | = x$ becomes a true statement whenever $x$ is replaced by zero or a positive number. The sentence $|\ x\ | = x$ becomes a false statement whenever $x$ is replaced by a negative number.

When $x$ is replaced by $^{-}5$, $|\ x\ | = x$ becomes

$$|\ {}^{-}5\ | = {}^{-}5$$

This numerical statement is false.

Similarly $|\ {}^{-}19\ | = {}^{-}19$ is a false statement obtained when $x$ is replaced by $^{-}19$ in $|\ x\ | = x$.

$|\ x\ | = x$ becomes true only in those cases where $x$ is replaced by zero or a positive real number.

When the absolute value of a negative number is taken then the result is its opposite. $|\,^{-}5\,| = 5$ and $|\,^{-}19\,| = 19$ are true statements. To state the fact that the absolute value of a negative number is its opposite the open sentence $|\,x\,| = \,^{-}x$ is used. This open sentence becomes a true statement for all negative number replacements of $x$.

$$|\,x\,| = \,^{-}x$$

$$|\,(^{-}5)\,| = \,^{-}(^{-}5) = 5$$

$$|\,(^{-}19)\,| = \,^{-}(^{-}19) = 19$$

Whenever the absolute value of an open expression is desired two possibilities must be considered. If the quantity inside the absolute value bars is positive then the absolute value will be equal to that quantity. When the quantity inside the absolute value bars is negative then the absolute value will be equal to the opposite of that quantity. These two cases of absolute value are shown by the following two open sentences.

$$|\,x\,| = x \qquad \text{when} \qquad x > 0$$

$$|\,x\,| = \,^{-}x \qquad \text{when} \qquad x < 0$$

Suppose $x$ is to be replaced by a real number greater than 10 in $|\,x - 5\,|$. In such a case $(x - 5)$ will always be positive and therefore $|\,x - 5\,| = x - 5$ will be a true statement because the absolute value of a positive number is that same positive number.

But suppose $x$ is to be replaced by a number less than 5 in $|\,x - 5\,|$. Then $(x - 5)$ would always be negative and $|\,x - 5\,| = \,^{-}(x - 5)$ will become a true statement because the absolute value of a negative number is its opposite.

If $x > 6$ is a true statement for some replacement of $x$ then $|\,x - 6\,|$ is equivalent to $x - 6$ because $(x - 6)$ is positive whenever $x > 6$.

If $x < 3$ is a true statement for some replacement of $x$ then $|\,x - 3\,|$ is equivalent to $^{-}(x - 3)$ because $(x - 3)$ is negative whenever $x < 3$.

EXERCISE 9.3.1

1. If $x < 5$ then $|\,x - 5\,| = $ _____.
2. If $x > \,^{-}2$ then $|\,x + 2\,| = $ _____.
3. If $x < \,^{-}6$ then $|\,x + 6\,| = $ _____.
4. If $x < 5$ then $|\,5 - x\,| = $ _____.
5. If $x > 8$ then $|\,x - 8\,| = $ _____.
6. If $x < 3$ then $|\,x - 3\,| = $ _____.
7. If $x > 5$ then $|\,x - 5\,| = $ _____.
8. If $x < \,^{-}9$ then $|\,x + 9\,| = $ _____.

9.  If $x > \dfrac{5}{2}$  then  $|\,2x - 5\,| =$ _____.

10. If $x < \dfrac{3}{4}$  then  $|\,4x - 3\,| =$ _____.

11. If $x < 7$  then  $|\,x - 7\,| =$ _____.
12. If $x > {}^{-}2$  then  $|\,x + 2\,| =$ _____.
13. If $x < {}^{-}3$  then  $|\,x + 3\,| =$ _____.
14. If $x > 5$  then  $|\,x - 5\,| =$ _____.
15. If $x < {}^{-}7$  then  $|\,x + 7\,| =$ _____.
16. If $x > 1$  then  $|\,x - 1\,| =$ _____.
17. If $x < {}^{-}5$  then  $|\,x + 5\,| =$ _____.

18. If $x > \dfrac{3}{2}$  then  $|\,2x - 3\,| =$ _____.

19. If $x < \dfrac{{}^{-}7}{3}$  then  $|\,3x + 7\,| =$ _____.

20. If $x > \dfrac{{}^{-}3}{5}$  then  $|\,5x + 3\,| =$ _____.

## 9.4  Absolute Value Equations and Two-Point Circles

In figure 9.1 a circle is shown which has its center on the number line. In what way does the size of the circle and the location of its center affect the two points where the circle intersects the number line?

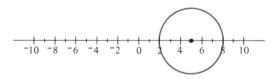

FIGURE 9.1

Because the radius of the circle is 3 and its center is at 5, the circle of figure 9.1 intersects the number line at 2 and 8. In what way is the following equation related to the circle of radius 3 which has its center at 5?

$$|\,x - 5\,| = 3$$

If you answered the preceding question by saying that $\{2, 8\}$ is the truth set of $|x - 5| = 3$ and these elements also represent the intersection of the circle with the number line then you are completely correct.

$$|x - 5| = 3$$

intersection    center    radius
point(s)

In figure 9.2 is shown a circle of radius 4 with its center on the number line. How is the circle of figure 9.2 related to the following equation?

$$|x - {}^-3| = 4$$

$\{^-7, 1\}$ is the truth set of $|x - {}^-3| = 4$ and these same elements are the points of intersection for the circle and the number line shown in figure 9.2.

$$|x - {}^-3| = 4$$

Intersection    center    radius
point(s)

Every equation of the form $|x - a| = b$ can be visualized as describing a circle which has its center on the real number line.

$$|x - a| = b$$

center    radius

The number replacing $a$ will indicate the center of the circle. The number replacing $b$ will be the radius and, therefore, must be positive if the circle is really to exist. Finally, there will be two replacements for $x$ that will make $|x - a| = b$ and $b > 0$ a true statement. These two replacements for $x$ will be the points of intersection of the circle with the number line.

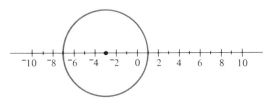

FIGURE 9.2

EXERCISE 9.4.1

1. Each circle is related to an equation of the form $|x - a| = b$ that shows the number relationship between the center of the

circle, its radius, and its intersection points with the number line. Write the correct absolute value equation associated with each circle.

(a)

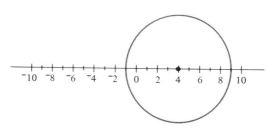

(b)

(c)

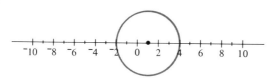

(d)

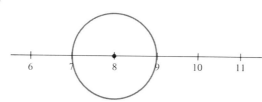

(e)

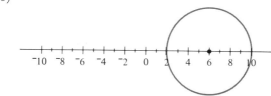

2. Draw a circle on the number line illustrating the relationship shown by each equation.

(a) $\mid x - 4 \mid = 7$
(b) $\mid x - 2 \mid = 1$
(c) $\mid x - {}^-2 \mid = 5$
(d) $\mid x - 3 \mid = 2$
(e) $\mid x - {}^-5 \mid = 3$
(f) $\mid x - 6 \mid = 3$
(g) $\mid x + 1 \mid = 9$    (Be careful!)
(h) $\mid x - 2 \mid = 6$
(i) $\mid x + 7 \mid = 4$
(j) $\mid x - 5 \mid = 1$

## 9.5  Distance

The reason that equations such as $\mid x - 4 \mid = 3$ and $\mid x - {}^-2 \mid = 5$ can be associated with circles intersecting the number line is the fact that every expression of the form $\mid x - y \mid$ can be considered to be the distance between $x$ and $y$.

For example, $\mid 5 - 3 \mid$ is the distance between 5 and 3. The absolute value bars in $\mid 5 - 3 \mid$ are unnecessary since $(5 - 3)$ is positive, but had the expression $\mid 3 - 5 \mid$ been used as the distance between 3 and 5 the use of the absolute value symbols insures that the distance will be positive despite the fact that $(3 - 5)$ is negative.

The use of absolute value in the expression $\mid x - y \mid$ insures that the result will be non-negative. Regardless of whether $x$ or $y$ is replaced by the greater number, $\mid x - y \mid$ will be the positive difference between the two numbers and, therefore, the distance from $x$ to $y$.

In the equation $\mid x - 3 \mid = 4$ the truth set will contain those numbers which are at a distance of 4 from 3. Since 7 and $^-1$ are both 4 from 3 then $\{7, {}^-1\}$ is the truth set of $\mid x - 3 \mid = 4$.

For the equation $\mid x + 5 \mid = 2$ it is best to write it as $\mid x - {}^-5 \mid = 2$ because the idea of distance requires the difference between two numbers. The truth set of $\mid x - {}^-5 \mid = 2$ will contain $^-7$ and $^-3$ because these are the two points at distance 2 from $^-5$.

EXERCISE 9.5.1

Find truth sets for:

1. $\mid x - 4 \mid = 1$          4. $\mid x - {}^-6 \mid = 2$
2. $\mid x - 8 \mid = 3$          5. $\mid x + 3 \mid = 3$
3. $\mid x - {}^-2 \mid = 5$      6. $\mid x + 4 \mid = 1$

| | | | |
|---|---|---|---|
| 7. | $\mid x - 7 \mid = 9$ | 14. | $\mid x - 18 \mid = 5$ |
| 8. | $\mid x + 1 \mid = 6$ | 15. | $\mid x - 14 \mid = 21$ |
| 9. | $\mid x - 11 \mid = 3$ | 16. | $\mid x + 25 \mid = 15$ |
| 10. | $\mid x - 12 \mid = 6$ | 17. | $\mid x - 31 \mid = 9$ |
| 11. | $\mid x + 13 \mid = 11$ | 18. | $\mid x - 24 \mid = 24$ |
| 12. | $\mid x + 9 \mid = 15$ | 19. | $\mid x + 13 \mid = 26$ |
| 13. | $\mid x + 20 \mid = 1$ | 20. | $\mid x + 45 \mid = 6$ |

## 9.6 Absolute Value Inequalities

In figure 9.3 is shown a circle of radius 4 with its center at 1. The circle of figure 9.3 might be considered as separating the number line into three sets of points: (1) The points of the number line inside the circle which are between $^-3$ and 5, (2) those points of the circle that are also on the number line which are $^-3$ and 5, and (3) those points of the number line that are outside the circle which are all those points to the right of 5 or to the left of $^-3$.

The truth set of $\mid x - 1 \mid = 4$ contains the two points of figure 9.3 that are on the circle, 5 and $^-3$, because these points are the only ones which are at distance 4 from 1. What open sentence might describe the set of all points on the number line inside the circle? Since these points are to be inside the circle and it has radius 4, a correct answer for the question is $\mid x - 1 \mid < 4$.

What open sentence might describe those points on the number line outside the circle? The radius is 4 and since these points are outside the circle, a correct answer could be $\mid x - 1 \mid > 4$.

In figure 9.4 a circle of radius 5 and center $^-2$ is shown on the number line. Again, the circle has separated the number line into three sets of points: (1) Those points inside the circle, (2) those points on the circle, and (3) those points outside the circle.

An open sentence to describe the points inside the circle is $\mid x - {}^-2 \mid < 5$. This is because the open expression $\mid x - {}^-2 \mid$ denotes the distance between $x$ and $^-2$ and those points inside the circle must have a distance from $^-2$ that is less than 5.

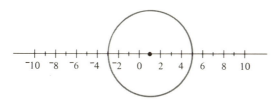

FIGURE 9.3

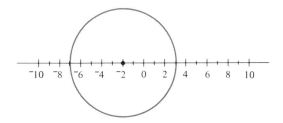

FIGURE 9.4

An open sentence to describe the points on the circle is $|x - {}^-2| = 5$.
An open sentence to describe the points outside the circle is $|x - {}^-2| > 5$ because the points outside the circle must have a distance from ${}^-2$ that is greater than 5.

EXERCISE 9.6.1

In each case write an open sentence of the form $|x - a| < b$, $|x - a| = b$, or $|x - a| > b$ to describe the set of points on the number line given by:

1. The points on the circle of radius 5 and center 3.
2. The points inside the circle of radius 3 and center ${}^-2$.
3. The points outside the circle of radius 2 and center 0.
4. The points inside the circle of radius 7 and center 7.
5. The points outside the circle of radius 4 and center ${}^-6$.
6. The points on the circle of radius 6 and center 4.
7. The points outside the circle of radius 5 and center ${}^-3$.
8. The points inside the circle of radius 8 and center ${}^-1$.
9. The points outside the circle of radius 6 and center ${}^-1$.
10. The points inside the circle of radius 4 and center 5.

## 9.7  Graphing Truth Sets of Absolute Value Inequalities

Once the truth set of an equation of the form $|x - a| = b$ is pictured as the intersection of the number line with a circle of radius $b$ and center $a$, then it is relatively easy to graph inequalities such as $|x - 4| < 7$ and $|x - {}^-3| > 4$.

The inequality $|x - 4| < 7$ can be pictured as that portion of the number line inside a circle of radius 7 and center 4.

The graph of $|x - 4| < 7$ is shown in figure 9.5.

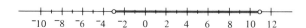

FIGURE 9.5

The inequality $\mid x - {}^-3 \mid > 4$ can be pictured as that portion of the number line outside the circle of radius 4 and center $^-3$.

The graph of $\mid x - {}^-3 \mid > 4$ is shown below in figure 9.6.

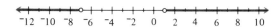

FIGURE 9.6

EXERCISE 9.7.1

Graph each inequality on a number line.

| | | | |
|---|---|---|---|
| 1. | $\mid x - 6 \mid > 2$ | 11. | $\mid x + 5 \mid > 9$ |
| 2. | $\mid x - {}^-2 \mid < 4$ | 12. | $\mid x - 1 \mid < 6$ |
| 3. | $\mid x - 2 \mid > 5$ | 13. | $\mid x + 3 \mid < 7$ |
| 4. | $\mid x - {}^-4 \mid > 2$ | 14. | $\mid x + 8 \mid < 6$ |
| 5. | $\mid x - 5 \mid > 5$ | 15. | $\mid x - 13 \mid > 2$ |
| 6. | $\mid x + 3 \mid < 4$ | 16. | $\mid x - 5 \mid > 7$ |
| 7. | $\mid x + 7 \mid > 1$ | 17. | $\mid x + 1 \mid > 3$ |
| 8. | $\mid x - 8 \mid > 3$ | 18. | $\mid x - 9 \mid < 5$ |
| 9. | $\mid x + 4 \mid < 4$ | 19. | $\mid x + 6 \mid < 6$ |
| 10. | $\mid x - 5 \mid < 2$ | 20. | $\mid x - 4 \mid > 7$ |

## 9.8  Solving Absolute Value Open Sentences Algebraically

We have seen how to solve graphically equations and inequalities such as $\mid x + 4 \mid = 9$ and $\mid x - 6 \mid < 4$. In this section another method will be shown for solving open sentences involving absolute values. This new method is applicable to all the open sentences already solved in this chapter as well as more complex ones such as $\mid 2x - 3 \mid > 5$ and $\mid x + 4 \mid + \mid 3x - 2 \mid = \mid x - 2 \mid$.

The absolute value of every real number is non-negative. Because of this, absolute value is often defined by the following pair of equalities.

$$\mid x \mid = x, \qquad \text{whenever } x \text{ is non-negative}$$

$$\mid x \mid = {}^-x, \qquad \text{whenever } x \text{ is negative}$$

The first equality simply claims that the absolute value of a positive number or zero is that same number. The second equality claims that the absolute value of a negative number is its opposite. When $x$ is replaced by a negative number then $^-x$ will be positive.

For the equation $|x - 5| = 8$ some replacements of $x$ will cause $(x - 5)$ to be non-negative while others will cause $(x - 5)$ to be negative. We solve $|x - 5| = 8$ using two cases: (1) When $(x - 5)$ is non-negative then $|x - 5| = x - 5$, (2) when $(x - 5)$ is negative $|x - 5| = {}^-(x - 5)$.

$$|x - 5| = 8$$

(1)  If $x - 5 \geq 0$ ($\geq$ means greater than or equal to)

$$|x - 5| = 8$$

$$x - 5 = 8$$

$\{13\}$ is the truth set.

Notice that when $x = 13$, $(x - 5)$ is positive.

(2)  If $x - 5 < 0$

$$|x - 5| = 8$$

$$^-(x - 5) = 8$$

$$^-x + 5 = 8$$

$$^-x = 3$$

$\{^-3\}$ is the truth set

Notice that when $x = {}^-3$, $(x - 5)$ is negative.
The truth set of $|x - 5| = 8$ is the union of $\{13\}$ and $\{^-3\}$, $\{13, {}^-3\}$.

To solve the inequality $|2x - 3| < 4$ we again consider the two possibilities, (1) $(2x - 3) \geq 0$, and (2) $(2x - 3) < 0$.

$$|2x - 3| < 4$$

(1)  If $(2x - 3) \geq 0$

$$|2x - 3| < 4$$

$$2x - 3 < 4$$

$$2x < 7$$

$$x < \frac{7}{2}$$

(2)   If $(2x - 3) < 0$

$| 2x - 3 | < 4$

$^{-}(2x - 3) < 4$

$^{-}2x + 3 < 4$

$^{-}2x < 1$

$$x > \frac{^{-}1}{2}$$

Had we been graphing the truth set of $| 2x - 3 | < 4$ it should have been obvious that the graph would be a line segment because it is the portion of

the line inside a circle. Hence, the inequalities $x < \frac{7}{2}$ and $x > \frac{^{-}1}{2}$ must *both*

be satisfied for any solution of $| 2x - 3 | < 4$. The notation $\frac{^{-}1}{2} < x < \frac{7}{2}$ is

used to show that *both* $x > \frac{^{-}1}{2}$ and $x < \frac{7}{2}$ must be satisfied for any solution

of $| 2x - 3 | < 4$. We say that $\frac{^{-}1}{2} < x < \frac{7}{2}$ is an equivalent open sentence

to the absolute value inequality $| 2x - 3 | < 4$.

To solve $| 3x + 2 | > 5$ we again consider the two cases: $(3x + 2) \geq 0$ and $(3x + 2) < 0$.

$$| 3x + 2 | > 5$$

(1)   If   $(3x + 2) \geq 0$

$| 3x + 2 | > 5$

$3x + 2 > 5$

$3x > 3$

$x > 1$

(2)   If   $(3x + 2) < 0$

$|\, 3x + 2 \,| > 5$

$-(3x + 2) > 5$

$-3x - 2 > 5$

$-3x > 7$

$x < \dfrac{-7}{3}$

The graph of $|\, 3x + 2 \,| > 5$ would consist of those portions of the number line that are outside a circle and this knowledge is applied to the inequalities

$x > 1$ and $x < \dfrac{-7}{3}$. Just as the graph of the number line outside the circle

consists of those points to the right of the circle *or* those points to the left of the circle, so too can we claim that when *either* inequality, $x > 1$ or

$x < \dfrac{-7}{3}$, is satisfied then $|\, 3x + 2 \,| > 5$ will become a true statement.

We say that $x > 1$ or $x < \dfrac{-7}{3}$ is an equivalent open sentence to the absolute

value inequality $|\, 3x + 2 \,| > 5$.

EXERCISE 9.8.1

Solve the following open sentences algebraically by considering the two cases of non-negative and negative quantities within the absolute value bars.

1.   Find truth sets for:
   (a)  $|\, x - 5 \,| = 6$
   (b)  $|\, x + 2 \,| = 4$
   (c)  $|\, 2x + 1 \,| = 9$
   (d)  $|\, 3x - 7 \,| = 2$
   (e)  $|\, 5x + 1 \,| = 6$
   (f)  $|\, 3x - 7 \,| = 4$
   (g)  $|\, 6x - 5 \,| = 10$
   (h)  $|\, 5x - 3 \,| = 7$
   (i)  $|\, 2x + 5 \,| = 3$
   (j)  $|\, 4x - 1 \,| = 9$

2.  Find equivalent open sentences using inequalities for:
    (a)  $|x - 3| > 7$
    (b)  $|x + 4| < 6$
    (c)  $|x - 8| > 1$
    (d)  $|x - 5| < {}^-3$     (Think!)
    (e)  $|3x - 2| > 7$
    (f)  $|2x + 5| > 9$
    (g)  $|4x + 7| < 3$
    (h)  $|4 - 3x| > 5$
    (i)  $|2x + 1| > {}^-2$
    (j)  $|3x + 1| < 2$
    (k)  $|5x - 2| > 3$
    (l)  $|2 - 4x| < 2$

# 10 RELATIONS AND FUNCTIONS

## 10.1 Relations

$\{(6, ^-1), (^-4, \sqrt{2}), (5, 7), (5, \pi)\}$ is a set and every element of the set is an ordered pair of real numbers. Whenever a set has only ordered pairs as its elements then we call the set a *relation*.

$\{(x, y) \mid 3x + y = 7\}$ is a method of describing a set of ordered pairs. It is read as "the set of all ordered pairs $(x, y)$ that are solutions of $3x + y = 7$." $(2, 1)$ is an element of the set because $3 \cdot 2 + 1 = 7$ is a true statement. $(3, 5)$ is not an element of the set because $3 \cdot 3 + 5 = 7$ is a false statement. Because every element of the set $\{(x, y) \mid 3x + y = 7\}$ is an ordered pair of real numbers we call the set a *relation*.

$\{(x, y) \mid x^2 + x - 4 = y\}$ describes the set of all ordered pairs $(x, y)$ that are solutions of $x^2 + x - 4 = y$. $(2, 2)$ is an element of the set because $2^2 + 2 - 4 = 2$ is a true statement. $(3, 5)$ is not an element of the set because $3^2 + 3 - 4 = 5$ is a false statement. Since every element of $\{(x, y) \mid x^2 + x - 4 = y\}$ is an ordered pair of real numbers we call the set a *relation*.

EXERCISE 10.1.1

1.  Which of the following sets are relations?
    (a) $\{(2, 7), (4, ^-1), (2, 6), (1, ^-3)\}$
    (b) $\{(5, 3), (6, 1), (6, 5), (4, ^-2), (9, 4)\}$
    (c) $\{2, 3, 5, 7, 9\}$
    (d) $\{(x, y) \mid x - 2y = 3\}$
    (e) $\{(x, y) \mid x^2 + 5x = y - 2\}$
    (f) $\{x \mid 2x = 13\}$
    (g) $\{(x, y) \mid 2x + 5y = 6\}$
    (h) $\left\{ (x, y) \mid \dfrac{3}{x - 2} = \dfrac{4}{y - 6} \right\}$
    (i) $\{(x, y) \mid \sqrt{x - 2} = y\}$
    (j) $\{(x, y) \mid \mid 3x + 5 \mid = y\}$

2.  Which of the ordered pairs are elements for each relation?
    (a) $(3, 6), (2, 5), (^-1, ^-4)$   for   $\{(x, y) \mid 5x - 2y = 3\}$
    (b) $(6, 15), (^-3, ^-3), (^-1, 2)$   for   $\{(x, y) \mid 3x + 6 = y\}$
    (c) $(6, 12), (3, 2), (^-1, 2)$   for   $\{(x, y) \mid x^2 - 3x - 2 = y\}$
    (d) $(2, 11), (^-6, 11), (0, 11)$   for   $\{(x, y) \mid x^2 + 4x - 1 = y\}$
    (e) $(3, 0), (^-5, 4), (5, ^-4)$   for   $\{(x, y) \mid \sqrt{x^2 - 9} = y\}$
    (f) $(4, \sqrt{5}), (^-11, 5), (^-23, 7)$   for   $\{(x, y) \mid \sqrt{3 - 2x} = y\}$
    (g) $\left(12, \dfrac{1}{4}\right), (0, 0), (^-3, ^-1)$   for   $\left\{ (x, y) \mid \dfrac{3}{x} = y \right\}$
    (h) $(1, 3), (3, 3), (13, 23)$   for   $\left\{ (x, y) \mid \dfrac{2}{x - 3} = \dfrac{5}{y + 2} \right\}$
    (i) $(^-2, 4), (2, 4), (^-10, 4)$   for   $\{(x, y) \mid \mid x + 6 \mid = y\}$
    (j) $(1, 2), (4, 5), (^-2, 10)$   for   $\{(x, y) \mid \mid 4x - 3 \mid > y\}$

## 10.2   Domain and Range

An ordered pair consists of two members or components. In the ordered pair $(2, 6)$ the 2 is called the first component and the 6 is called the second component.

The set $\{(x, y) \mid 3x - y < 7\}$ is a set of ordered pairs or a relation. Those real numbers that serve as replacements for $x$ will be first components of ordered pairs. Replacements for $y$ will be second components.

Every relation or set of ordered pairs determines two other sets of real numbers. If all of the first components are used as elements of a set then we call that set the *domain* of the relation. The set of all second components is called the *range* of the relation.

$\{(2, 4), (3, 7), (4, 8), (5, 8), (6, 7)\}$ is a relation because every element of the set is an ordered pair. The first components of these ordered pairs are 2, 3, 4, 5, and 6 and, therefore, the *domain* of the relation is $\{2, 3, 4, 5, 6\}$. The second components of the relation are 4, 7, and 8; therefore, the *range* of the relation is $\{4, 7, 8\}$.

Whenever a relation consists of only a few ordered pairs the domain and range are easily found in the manner of the preceding paragraph. When a relation such as $\left\{(x, y) \,\middle|\, x^2 = \dfrac{5}{y - 3}\right\}$ is under consideration, the determination of the domain and range requires deeper study.

To find the domain and range of the relation $\left\{(x, y) \,\middle|\, x^2 = \dfrac{5}{y - 3}\right\}$ we inspect the equality $x^2 = \dfrac{5}{y - 3}$.

The domain of $\left\{(x, y) \,\middle|\, x^2 = \dfrac{5}{y - 3}\right\}$ consists of all those real numbers that can be used as replacements for $x$ in solutions for $x^2 = \dfrac{5}{y - 3}$. We will know the domain of the relation if we determine those real numbers that must be excluded as possible replacements for $x$.

There are two questions to answer:

(1)   $x$ appears only as $x^2$ in $x^2 = \dfrac{5}{y - 3}$. Is there any real number that can not be squared? No. So far any real number can replace $x$.

(2)   Since $x^2$ is equal to $\dfrac{5}{y - 3}$, are there any real numbers that can not be obtained through suitable replacements for $y$ in $\dfrac{5}{y - 3}$?

In this case the answer is "Yes." No replacement for $y$ can make $\dfrac{5}{y - 3}$ equal to zero, and therefore, neither $x^2$ nor $x$ can be zero.

The answers to the preceding two questions lead to the conclusion that the domain of $\left\{ (x, y) \left| x^2 = \dfrac{5}{y - 3} \right. \right\}$ is the set of all real numbers except zero.

Now the range of the relation is found by asking similar questions for real number replacements of $y$.

(1)   In the equation $x^2 = \dfrac{5}{y - 3}$ , $y$ appears in the expression $\dfrac{5}{y - 3}$ . Is there any real number that can not serve as a replacement for $y$? The answer is "Yes." 3 can not replace $y$ because such a replacement would result in a zero denominator.

(2)   Since $\dfrac{5}{y - 3}$ is equal to $x^2$, are there any real numbers that can not be obtained through suitable replacements of $x$ in $x^2$? The answer again is "Yes." No replacement for $x$ will result in a negative number for $x^2$ and therefore $\dfrac{5}{y - 3}$ must never be negative.

This leads to the conclusion that $y$ must be greater than 3.

The answers to the two preceding questions lead to the range of $\left\{ (x, y) \left| x^2 = \dfrac{5}{y - 3} \right. \right\}$ as being all real numbers greater than 3.

The complications in finding the domain and range of $\left\{ (x, y) \left| x^2 = \dfrac{5}{y-3} \right. \right\}$ are caused by the exponent 2 and the use of $y$ in a denominator. Two other difficulties often arise in finding domain and range sets. They are the use of absolute value and radical signs. Exponents, variables in denominators, absolute values, and radical signs should be studied closely whenever they are present in a relationship. Generally they will affect either the domain or the range of the relation.

EXERCISE 10.2.1

1.   For the equation $x^2 + 5 = y$ the only replacements for $y$ that can make true statements are real numbers greater than or equal to 5. Why?

2.  For the equation $|y - 6| = x$ the only replacements for $x$ that can make true statements are non-negative real numbers. Why?
3.  For the equation $\sqrt{9 - x} = y$ replacements for $x$ must be less than or equal to 9 and replacements for $y$ must be non-negative. Why?
4.  Find the domain and range of each relation.
    (a)  $\{(4, 7), (^-2, 3), (5, 2), (^-2, 6)\}$
    (b)  $\{(4, ^-1), (2, ^-1), (9, ^-1), (6, ^-1), (13, ^-1)\}$
    (c)  $\{(x, y) \mid x + y = 7\}$
    (d)  $\{(x, y) \mid x^2 - 6 = y\}$
    (e)  $\left\{(x, y) \,\middle|\, x = \dfrac{1}{y - 2}\right\}$
    (f)  $\{(x, y) \mid \sqrt{4 - x} = y\}$
    (g)  $\{(x, y) \mid 2 + |x| = y\}$

## 10.3  Graphing Relations

Every relation is a set of ordered pairs. As such it can be shown graphically using a pair of axes like those shown in figure 10.1.

Shown in Figure 10.1 is the graph of the relation

$$\{(2, 4), (^-3, 5), (0, ^-4), (0, 7), (^-6, ^-1)\}.$$

Because this relation contains only five ordered pairs, its graph consists of only five points.

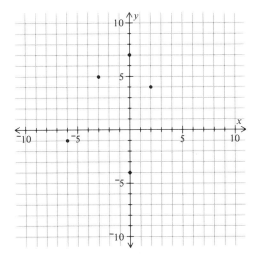

FIGURE 10.1

$\{(x, y) \mid 3x - y = 6\}$ is a relation which has no limit to its elements. You should recognize that $3x - y = 6$ is like the equations of chapter 5 and, therefore, the graph of the relation is the same as the graph of $3x - y = 6$. The graph of $\{(x, y) \mid 3x - y = 6\}$ is shown in figure 10.2.

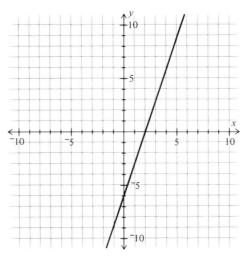

FIGURE 10.2

The graph of $\{(x, y) \mid x^2 + y^2 = 16\}$ is a circle of radius 4 with its center at $(0, 0)$. The graph of $\{(x, y) \mid x^2 + y^2 = 16\}$ is shown in figure 10.3.

To graph a relation is to pictorially represent all of its ordered pair elements. In the case of relations like $\{(^-4, 7), (^-2, 6), (1, 2)\}$ this is easy

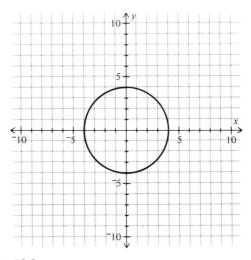

FIGURE 10.3

because there are so few points to show. In the case of relations like $\{(x, y) \mid 2x + y = 4\}$ we have already seen that such equations always have straight line graphs and methods were developed in chapter 5 for graphing them. In the case of relations such as $\{(x, y) \mid |x - 4| = y\}$ and $\{(x, y) \mid x^2 - 6 = y\}$ we shall employ the following procedure:

(1) Find ordered pair solutions for $(^-10, \underline{\phantom{x}})$, $(^-8, \underline{\phantom{x}})$, $(^-6, \underline{\phantom{x}})$, $(^-4, \underline{\phantom{x}})$, $(^-2, \underline{\phantom{x}})$, $(0, \underline{\phantom{x}})$, $(2, \underline{\phantom{x}})$, $(4, \underline{\phantom{x}})$, $(6, \underline{\phantom{x}})$, $(8, \underline{\phantom{x}})$, $(10, \underline{\phantom{x}})$.

(2) Assuming that the graph is fairly consistent between these eleven points, connect them from left to right with a smooth line.

Let us graph $\{(x, y) \mid |x - 4| = y\}$ using the procedure given above.

To complete the solution for $(^-10, \underline{\phantom{x}})$ we replace $x$ by $^-10$ in $|x - 4| = y$ and find $y$.

$$|x - 4| = y$$

$$|^-10 - 4| = y$$

$$|^-14| = y$$

$$14 = y$$

$(^-10, 14)$ is a solution of $\{(x, y) \mid |x - 4| = y\}$.

Similarly we find $(^-8, 12)$, $(^-6, 10)$, $(^-4, 8)$, $(^-2, 6)$, $(0, 4)$, $(2, 2)$, $(4, 0)$, $(6, 2)$, $(8, 4)$, $(10, 6)$ are other elements of the relation. Graphing these eleven solutions and connecting the points produces the graph below. Figure 10.4 is the graph of $\{(x, y) \mid |x - 4| = y\}$.

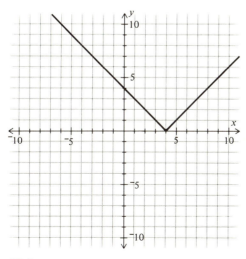

FIGURE 10.4

Graphing $\{(x, y) \mid x^2 + y^2 = 36\}$ using the procedure of finding various points and connecting them will lead us to a circle. To complete $(^-10, \_)$ as a solution for $x^2 + y^2 = 36$ we replace $x$ by $^-10$ and try to find $y$.

$$x^2 + y^2 = 36$$
$$(^-10)^2 + y^2 = 36$$
$$100 + y^2 = 36$$
$$y^2 = ^-64$$

There is no real number that when squared will give $^-64$. Hence there is no solution with a first component of $^-10$. A similar result is arrived at in trying to complete $(^-8, \_)$ as a solution of $x^2 + y^2 = 36$.

To complete $(^-6, \_)$:

$$x^2 + y^2 = 36$$
$$(^-6)^2 + y^2 = 36$$
$$36 + y^2 = 36$$
$$y^2 = 0$$

Hence $(^-6, 0)$ is a solution.

To complete $(^-4, \_)$:

$$x^2 + y^2 = 36$$
$$(^-4)^2 + y^2 = 36$$
$$16 + y^2 = 36$$
$$y^2 = 20$$
$$y = 2\sqrt{5} \quad \text{or} \quad y = ^-2\sqrt{5}$$

Hence both $(^-4, 2\sqrt{5})$ and $(^-4, ^-2\sqrt{5})$ are solutions of $x^2 + y^2 = 36$.

Similarly, $(^-2, 4\sqrt{2})$, $(^-2, ^-4\sqrt{2})$, $(0, 6)$, $(0, ^-6)$, $(2, 4\sqrt{2})$, $(2, ^-4\sqrt{2})$, $(4, 2\sqrt{5})$, $(4, ^-2\sqrt{5})$, and $(6, 0)$ are other solutions of $x^2 + y^2 = 36$. Graphing these points we find the arrangement shown in figure 10.5.

Drawing a smooth curve through the points of figure 10.5 we complete the graph of $\{(x, y) \mid x^2 + y^2 = 36\}$ in figure 10.6.

EXERCISE 10.3.1

Graph each relation.

1. $\{(4, 2), (^-6, 5), (^-2, ^-6), (3, ^-1), (5, 2)\}$
2. $\{(6, 1), (6, 2), (6, 3), (6, 4), (6, 5), (6, 6)\}$
3. $\{(x, y) \mid 2x + 3y = 12\}$
4. $\{(x, y) \mid x - 5y = 10\}$

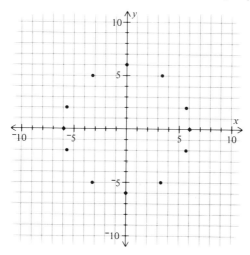

FIGURE 10.5

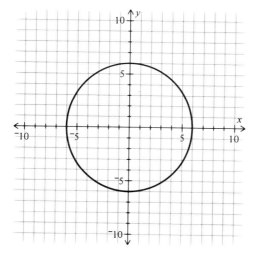

FIGURE 10.6

5. $\{(x, y) \mid x^2 - 4 = y\}$
6. $\{(x, y) \mid x^2 + 6x + 9 = y\}$
7. $\{(x, y) \mid x^2 - x - 12 = y\}$
8. $\{(x, y) \mid x^2 + 5x + 7 = y\}$
9. $\{(x, y) \mid \ \mid x - 3 \mid = y\}$
10. $\{(x, y) \mid \ \mid 2x + 1 \mid = y\}$
11. $\{(x, y) \mid \ \mid x \mid + \mid y \mid = 5\}$
12. $\{(x, y) \mid \ \mid x - 6 \mid = y - 2\}$
13. $\{(x, y) \mid x^2 + y^2 = 81\}$
14. $\{(x, y) \mid x^2 + y^2 = 25\}$

### 10.4  Functions

Any set of ordered pairs is a relation. In this section we explain the concept of function. A function is a special type of relation. Every function must be a set of ordered pairs because every function is a relation. Not every relation will be a function because a function is a special type of relation.

The criterion for determining that a relation is a function is:

A relation is a function when every domain element is used exactly once as the first component of an ordered pair.

Here is a relation that is a function:

$$\{(4, 7), (5, 6), (11, 7), (12, 7), (^-1, 2)\}$$

Notice:

(1)  There are five ordered pairs in the relation and five elements in the domain, $\{4, 5, 11, 12, ^-1\}$. Every element in the domain is used exactly once as a first component of an ordered pair.

(2)  If an element of the domain is selected then there is exactly one ordered pair that has it as its first component. For example, if 11 is selected as the domain component then the ordered pair (11, __) can be completed only by 7 to give (11, 7) as an ordered pair of the function.

(3)  The graph of the function is shown in figure 10.7. Any vertical line on this graph can not intersect more than one point of the function.

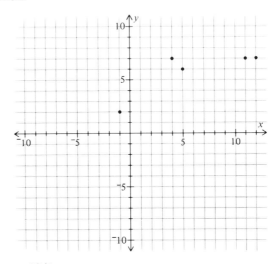

FIGURE  10.7

Here is a relation that is not a function:

$$\{(6, 3), (4, 7), (7, 8), (7, 1), (9, {}^-2)\}$$

Compare the following three statements with those made for the function graphed in figure 10.7.

(1)   There are five ordered pairs in the relation and only four elements in the domain, $\{6, 4, 7, 9\}$. There is some element of the domain that is used more than once as a first component.

(2)   If a domain element is selected there may be more than one ordered pair in the relation having it as first component. For example, if 7 is selected as the domain element then $(7, \_\_)$ can be completed in two ways, $(7, 8)$ or $(7, 1)$, to give elements of the relation.

(3)   The graph of the relation is shown in figure 10.8. There is a vertical line shown on the graph that intersects more than one point of the relation.

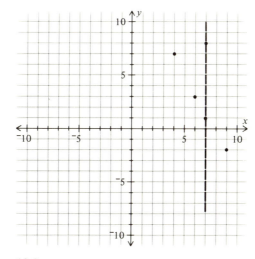

FIGURE  10.8

EXERCISE  10.4.1

1.   Determine whether each of the following is a function.

(a)  $\{(1, 3), (2, 7), (5, 3), (9, 7), (10, 1), (11, 2)\}$
(b)  $\{(2, 7), (3, 9), (4, 7), (4, 8), (5, 9)\}$
(c)  $\{(2, {}^-2), (2, 4), (2, 6), (2, 7)\}$

(d)  $\{(6, {}^-15)\}$
(e)  $\{(9, {}^-1), ({}^-2, 5), ({}^-4, 6), (0, 5), (6, {}^-1)\}$
(f)  $\{(6, 4), (2, 5), ({}^-2, 1), (5, 1)\}$

2.  The graph of a relation may be used to determine whether it is a function. If any vertical line intersects two or more points of the relation, the relation is not a function. Otherwise it is a function. Determine whether each of the following is the graph of a function.

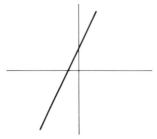

(a)     $\{(x,y) \mid 2x + 3 = y\}$

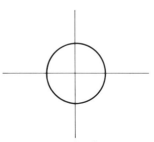

(b)     $\{(x,y) \mid x^2 + y^2 = 16\}$

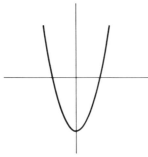

(c)     $\{(x,y) \mid x^2 - 9 = y\}$

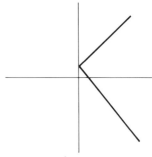

(d)     $\{(x,y) \mid |y - 2| = x\}$

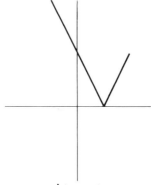

(e)     $\{(x,y) \mid |2x - 5| = y\}$

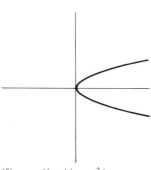

(f)     $\{(x,y) \mid x = y^2\}$

3.  For each relation below find all ordered pairs of the form
    (2, __) that are elements of the set. These equalities were
    especially chosen because if there is only one such ordered pair
    the relation is a function.

    (a)  (2, __)   $\{(x, y) \mid 2x + 3 = y\}$
    (b)  (2, __)   $\{(x, y) \mid x^2 + y^2 = 16\}$
    (c)  (2, __)   $\{(x, y) \mid x^2 - 9 = y\}$
    (d)  (2, __)   $\{(x, y) \mid \mid y - 2 \mid = x\}$
    (e)  (2, __)   $\{(x, y) \mid \mid 2x - 5 \mid = y\}$
    (f)  (2, __)   $\{(x, y) \mid x = y^2\}$

## 10.5   Quadratic Functions

Every relation of the form $\{(x, y) \mid ax + by = c\}$ where $a$, $b$, and
$c$ are real numbers and $b \neq 0$ is a function. Such functions are called linear
functions because the equation used to determine elements of the set is
a linear equation.

A second type of function is of the form $\{(x, y) \mid ax^2 + bx + c = y\}$
where $a$, $b$, and $c$ are real numbers and $a \neq 0$. This type of function is called
a quadratic function.

There are two ways to explain why every relation of the form
$\{(x, y) \mid ax^2 + bx + c = y\}$ is a function: (1) If $x$ is replaced by any real
number in $ax^2 + bx + c = y$ then the left side of the equality will become
strictly a numerical expression involving the multiplication and addition of
real numbers. Any replacement for $x$ will result in exactly one acceptable
replacement for $y$. (2) Whenever the solutions of an equation of the form
$ax^2 + bx + c = y$ are graphed then the result is a curve somewhat like
one of those shown in figure 10.9. Notice that each of these curves could be
intersected no more than one time by any vertical line.

A general idea of the graph of a quadratic function can be obtained by
(1) using the coefficient, $a$, of $x^2$ in $ax^2 + bx + c = y$, and (2) solving the
equation $ax^2 + bx + c = 0$ for its truth set.

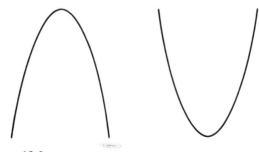

FIGURE 10.9

The coefficient of $x^2$ is used to indicate the direction and steepness of the curve. When the coefficient of $x^2$ is positive the curve always "opens up" like the one shown in figure 10.10.

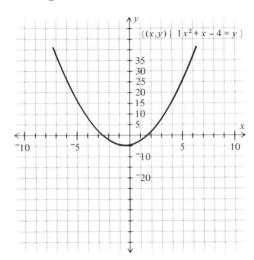

*

FIGURE 10.10

When the coefficient of $x^2$ is negative the curve always "opens down" like the one shown in figure 10.11.

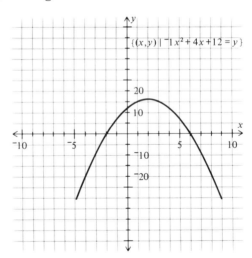

FIGURE 10.11

 * A different scale has been used on the $y$ axis to make the graph more manageable.

The absolute value of the coefficient of $x^2$ is important in determining the steepness of the curve. The greater the absolute value the steeper the curve. This fact is illustrated in figure 10.12 which shows the curves for

$3x^2 = y$ and $\dfrac{-1}{2}x^2 = y.$ | 3 | is greater than $\left| \dfrac{-1}{2} \right|$ and figure 10.12 shows

that $3x^2 = y$ has a steeper curve than $\dfrac{-1}{2}x^2 = y.$

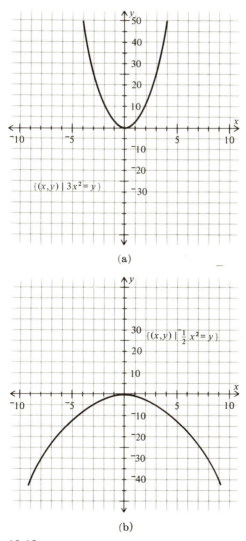

(a)

(b)

FIGURE 10.12

In figure 10.13 the graphs of two more quadratic functions are shown. Notice the manner in which the coefficient of $x^2$ affects the graphs: (1) When the coefficient is positive the graph opens up; when the coefficient is negative the graph opens down, and (2) the greater the absolute value of the coefficient of $x^2$, the steeper the curve.

In graphing a quadratic function like $\{(x, y) \mid 2x^2 - 7x + 3 = y\}$ the truth set of $2x^2 - 7x + 3 = 0$ is helpful because it gives the points where the graph crosses the $x$ axis and $y = 0$. Since $2x^2 - 7x + 3$ is factored

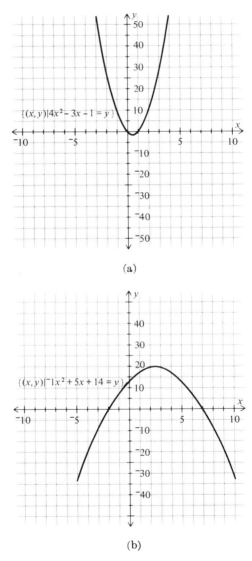

(a)

(b)

FIGURE 10.13

as $(2x - 1)(x - 3)$, the truth set of $2x^2 - 7x + 3 = 0$ is $\left\{\frac{1}{2}, 3\right\}$. This

means that the graph of the function crosses the $x$ axis at $\left(\frac{1}{2}, 0\right)$ and $(3, 0)$.

It also means that the curve changes direction midway between $\frac{1}{2}$ and $3$ or

at $\frac{7}{4}$. If $x$ is replaced by $\frac{7}{4}$ in the equality $2x^2 - 7x + 3 = y$, then the value

of $y$ at the "low" point of the graph will have been determined.

$$2x^2 - 7x + 3 = y$$

$$2\left(\frac{7}{4}\right)^2 - 7 \cdot \frac{7}{4} + 3 = y$$

$$\frac{49}{8} - \frac{49}{4} + 3 = y$$

$$\frac{-25}{8} = y$$

Graphing the three points $\left(\frac{1}{2}, 0\right)$, $(3, 0)$, and $\left(\frac{7}{4}, \frac{-25}{8}\right)$ and using the

information gained from the coefficient of $x^2$ in $2x^2 - 7x + 3 = y$, the
function is graphed as shown in figure 10.14.

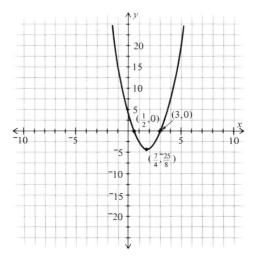

FIGURE 10.14

EXERCISE 10.5.1

For each quadratic function tell whether the graph "opens up" or "opens down." Also replace $y$ by zero and find where the graph crosses the $x$ axis.

1. $\{(x, y) \mid x^2 - 5x + 6 = y\}$
2. $\{(x, y) \mid 2x^2 - 9x + 9 = y\}$
3. $\{(x, y) \mid {}^-x^2 + 7x + 8 = y\}$
4. $\{(x, y) \mid {}^-3x^2 + 11x - 8 = y\}$
5. $\{(x, y) \mid x^2 + 10x + 16 = y\}$
6. $\{(x, y) \mid {}^-2x^2 - 3x + 5 = y\}$
7. $\{(x, y) \mid x^2 - 4x - 12 = y\}$
8. $\{(x, y) \mid 6x^2 - x - 1 = y\}$
9. $\{(x, y) \mid {}^-x^2 + 6x + 16 = y\}$
10. $\{(x, y) \mid x^2 + 4x - 21 = y\}$

### 10.6  Graphing Quadratic Functions

In the last section some suggestions for picturing the graphs of quadratic functions were given. The coefficient of $x^2$ can be used to determine the direction and steepness of the curve. Replacing $y$ by zero and solving the resultant quadratic equation provides the points, if any, where the graph crosses the $x$ axis. These types of graphing concepts can be extended to other, more sophisticated, methods, but they are beyond the intent of this text.

For our purposes we can continue to graph quadratic functions by plotting those solutions gained by replacing $x$ by the elements of $\{{}^-10, {}^-9, {}^-8, {}^-7, \ldots, 8, 9, 10\}$. For example, we graph $\{(x, y) \mid x^2 - 6x - 16 = y\}$ by completing the ordered pairs $({}^-10, \underline{\hspace{0.5cm}})$, $({}^-9, \underline{\hspace{0.5cm}})$, $({}^-8, \underline{\hspace{0.5cm}})$, etc. The graph of $\{(x, y) \mid x^2 - 6x - 16 = y\}$ is shown in figure 10.15.

$\{(x, y) \mid 2x^2 - 5x - 14 > y\}$ is not a function because the inequality used to determine ordered pair elements of the set allows first components to appear more than one time. The graph of the set can be found by first graphing the function $\{(x, y) \mid 2x^2 - 5x - 14 = y\}$, then selecting a point either "inside" the curve or "outside" it and testing that ordered pair for membership in $\{(x, y) \mid 2x^2 - 5x - 14 > y\}$.

The graph of the function $\{(x, y) \mid 2x^2 - 5x - 14 = y\}$ is shown in figure 10.16.

Using the graph of $\{(x, y) \mid 2x^2 - 5x - 14 = y\}$ shown above, the graph of $\{(x, y) \mid 2x^2 - 5x - 14 > y\}$ is found by:

(1)  Select a point not on the curve. Suppose $(1, 0)$ is selected.

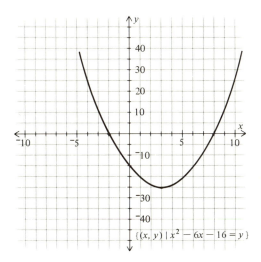

FIGURE 10.15

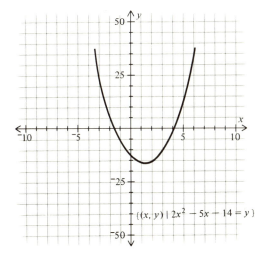

FIGURE 10.16

(2)   Test $(1, 0)$ to see if it is a solution of the inequality $2x^2 - 5x - 14 > y$.

$$2 \cdot 1^2 - 5 \cdot 1 - 14 > 0$$

$$2 - 5 - 14 > 0$$

$$^-17 > 0$$

(3)   $^-17 > 0$ is false. $(1, 0)$ and all other points located "inside" of the curve will make $2x^2 - 5x - 14 > y$ a false statement.

(4)   The graph of $\{(x, y) \mid 2x^2 - 5x - 14 > y\}$ consists of all points "outside" the curve as shown in figure 10.17.

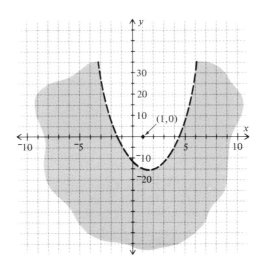

FIGURE  10.17

The graph of any relation of the form $\{(x, y) \mid ax^2 + bx + c > y\}$ or $\{(x, y) \mid ax^2 + bx + c < y\}$ will consist of all the points "inside" or "outside" the curve of the function $\{(x, y) \mid ax^2 + bx + c = y\}$. The relations are graphed by graphing the function and testing a point not on the curve. If the point tested makes the inequality true then all points on the same side of the curve are in the relation. If the point tested makes the inequality false then all points on the other side of the curve are in the relation.

EXERCISE 10.6.1

Graph:

1.   $\{(x, y) \mid x^2 + 4x - 12 = y\}$
2.   $\{(x, y) \mid {}^-x^2 - 3x + 18 = y\}$
3.   $\{(x, y) \mid x^2 - 5x - 6 > y\}$
4.   $\{(x, y) \mid x^2 + 4x - 32 < y\}$

5.  $\{(x, y) \mid {}^-x^2 + 2x - 1 > y\}$
6.  $\{(x, y) \mid x^2 + 3x + 1 < y\}$
7.  $\{(x, y) \mid 2x^2 - 5x + 2 > y\}$
8.  $\{(x, y) \mid {}^-3x^2 + x + 24 > y\}$
9.  $\{(x, y) \mid x^2 + x + 4 < y\}$
10. $\{(x, y) \mid {}^-x^2 - 3x - 3 > y\}$

## 10.7  Graphing Inequalities of the Forms $ax^2 + bx + c > 0$ and $ax^2 + bx + c < 0$

The open sentence $2x^2 - 7x - 9 > 0$ is an inequality in one variable and, therefore, the truth set can be graphed on a number line. The truth set of $2x^2 - 7x - 9 > 0$ is shown in figure 10.18.

According to the graph, any point to the right of $\dfrac{9}{2}$ or to the left of $^-1$

represents a number that will make $2x^2 - 7x - 9 > 0$ a true statement. Two questions might arise here: (1) How was the graph acquired? (2) What does this type of problem have to do with relations and functions which are sets of ordered pairs? The answer to this second question will also answer the first one.

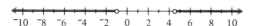

FIGURE 10.18

In what way are the open sentences $2x^2 - 7x - 9 > 0$ and $2x^2 - 7x - 9 = y$ related? Obviously the left members of the two open sentences are identical and comparing the two sentences leads to the conclusion that the truth set of $2x^2 - 7x - 9 > 0$ is the same as the first components of elements of $\{(x, y) \mid 2x^2 - 7x - 9 = y\}$ that have a second component greater than zero. Graphically then, we are talking about that portion of the graph of $\{(x, y) \mid 2x^2 - 7x - 9 = y\}$ that is above the $x$ axis; $y$ must be positive.

The graph of the function $\{(x, y) \mid 2x^2 - 7x - 9 = y\}$ is shown in

figure 10.19. It crosses the $x$ axis at $^-1$ and $\dfrac{9}{2}$ because these numbers are the

elements of the truth set of $2x^2 - 7x - 9 = 0$.

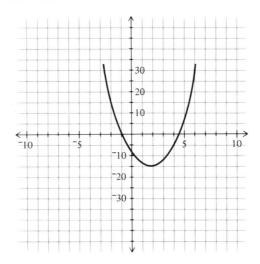

FIGURE 10.19

The graph of figure 10.19 is above the $x$ axis ($y$ is positive) to the right of $\frac{9}{2}$ and to the left of $^-1$. Therefore the graph of the truth set of $2x^2 - 7x - 9 > 0$ is the same portions of the $x$ axis as shown in figure 10.20.

FIGURE 10.20

Let us take another example and graph the truth set of $^-x^2 + 7x + 8 > 0$.

(1)  Visualize the graph of $\{(x, y) \mid ^-x^2 + 7x + 8 = y\}$. Since the coefficient of $x^2$ is $^-1$, the graph of the curve is going to "open down."

FIGURE 10.21

(2)   Find the points where the curve crosses the $x$ axis by solving $^-x^2 + 7x + 8 = 0$.

$$^-x^2 + 7x + 8 = 0$$

$$x^2 - 7x - 8 = 0$$

$$(x - 8)(x + 1) = 0$$

$$\{^-1, 8\}$$

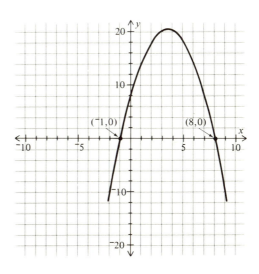

FIGURE 10.22

(3)   Since the truth set of $^-x^2 + 7x + 8 > 0$ will be those values of $x$ for which $y$ is positive in $\{(x, y) \mid ^-x^2 + 7x + 8 = y\}$, the portion of the curve that is important is that part above the $x$ axis.

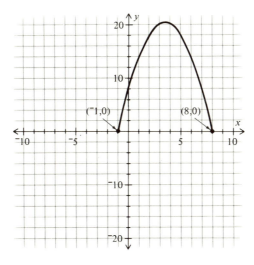

FIGURE 10.23

(4)    The values of $x$ related to that portion of the curve shown in figure 10.23 is the truth set of $^-x^2 + 7x + 8 > 0$ and its graph is shown in figure 10.24.

FIGURE 10.24

The crucial points to remember in the preceding two examples are

(1)    A visualization of the curve as opening "up" or "down" is important. It determines whether the truth set is to be two half-lines or one segment.

(2)    The endpoints of these half-lines or line segment are going to be the elements of the truth set of the quadratic equation obtained when $y = 0$.

One last example which is slightly different will be given here. To find the truth set of $x^2 + 2x + 6 < 0$:

(1)    Since $x^2$ has a coefficient of 1, the graph of the function $\{(x, y) \mid x^2 + 2x + 6 = y\}$ "opens up."

(2)    Solving $x^2 + 2x + 6 = 0$ for its truth set we find that there are no real number solutions. Consequently the graph must never cross the $x$ axis.

(3)    A sketch of the situation uncovered in (1) and (2) shows a graph like figure 10.25.

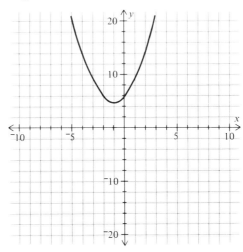

FIGURE 10.25

(4)  Since the truth set of $x^2 + 2x + 6 < 0$ is associated with all those points on the curve where $y$ is negative, we are forced to conclude that there are no such points. The truth set of $x^2 + 2x + 6 < 0$ is the empty set or { }.

Using the graph shown in figure 10.25, had the problem been to find the truth set of $x^2 + 2x + 6 > 0$ we would have to conclude that the truth set included all real numbers. The graph of the function is always above the $x$ axis and, therefore, $x^2 + 2x + 6$ is always positive for all real number replacements of $x$.

EXERCISE 10.7.1

Graph truth sets for:

1.  $x^2 + 6x - 27 > 0$          6.  $^-x^2 - 6x - 5 < 0$
2.  $x^2 - 3x - 4 < 0$          7.  $^-x^2 + 3x - 3 > 0$
3.  $x^2 + 9x + 18 > 0$          8.  $x^2 + 13x + 30 < 0$
4.  $x^2 - 5x + 8 < 0$          9.  $x^2 - 5x + 4 > 0$
5.  $^-x^2 + 2x + 8 > 0$          10.  $x^2 + 8x - 33 < 0$

# Index